NODES AND WEIGHTS OF QUADRATURE FORMULAS

UZLY I VESA KVADRATURNYKH FORMUL

УЗЛЫ И ВЕСА КВАДРАТУРНЫХ ФОРМУЛ

NODES AND WEIGHTS
OF
QUADRATURE FORMULAS

Sixteen-Place Tables

Aleksandr Semenovich Kronrod

Authorized translation from the Russian

CONSULTANTS BUREAU
NEW YORK
1965

First Printing – June 1965
Second Printing – April 1966

The original Russian text was published by
Nauka (Press of the Academy of Sciences of the USSR)
in Moscow in 1964.

Александр Семенович Кронрод
Узлы и веса квадратурных формул
(шестнадцатизначные таблицы)

Library of Congress Catalog Card Number 65-15002

Printed in the United States of America

PREFACE

The author has obtained formulas for the nodes and weights of improved quadratures. He gratefully acknowledges use of an idea by A. L. Brudno for the simple proofs of theorems 4 and 6.

The author extends his deep appreciation to a number of colleagues who helped him during the work on the actual computation of the weights and nodes of both the Gaussian and the improved quadratures with a large number of digits.

A. V. Uskov constructed a double-precision library routine. This routine played a large role in the intermediate stages even though the final computation of the tables was carried out with the aid of other routines. The same is also true of the library routine for working with integers and simple fractions which was constructed by A. A. Zhivotovskii and V. K. Pruss.

The burdensome, multidigit hand computations, which were necessary at the start of the work, were carried out by T. A. Muratova, I. Ya. Baranova, G. A. Zaitseva, and L. G. Bakylina.

The difficult and tedious work to typing the tables was done by Z. N. Zaletova; a large amount of work on punching out the program and on the digital computer was performed at the collective headed by A. A. Ganyushkina.

N. M. Kuznetsova aided the author greatly in various ways: in constructing a multiprecision library routine, in programming the computation of the roots of a function for numbers in multiprecision form, in setting up and debugging the computation programs for the nodes and weights of the Gaussian and the improved quadratures. The same is also true for the punch control programs for the tables.

Besides this, N. M. Kuznetsova carried out the work on punch control, on checking out all the tabulations, on correcting and final-checking the tables, all of which was laborious and required extreme care.

Finally, the deep thanks of the author go to the engineering-technical personnel of the electronic computer facility headed by Nikolaev Ivanovich Bessonov (now deceased). The writing of this book was made possible by the excellent work on the computer.

Institute of Theoretical and Experimental Physics
State Committee on the Use of Atomic Energy

A. S. Kronrod

CONTENTS

THEORETICAL ASPECTS

1. STATEMENT OF THE PROBLEM

We consider the linear quadratures for computing the integrals $\int_a^b f(x)\,dx$, i.e., the formulas having the form:

$$A\,(f) = \sum_{i=1}^{n} V_i\, f[a + (b-a)\, x_i].\qquad(1)$$

The numbers x_i are called the *nodes* and the numbers V_i the *weights* of quadrature A. We may of course assume $V_i \neq 0$, and in what follows we shall do so without stating it explicitly.

Quadratures of special types may prove to be advantageous for functions of special forms (for example, those decreasing exponentially with given indices, etc.). However, if we do not make any assumptions concerning the behavior of the integrand, the classical statement of the problem is: the quadrature formula of form (1) should accurately integrate polynomials of possibly very high degree when the number n of integration nodes is given. Gaussian quadratures are of such type (see, for example, A. N. Krylov, *Lectures on Approximate Computations*, Gostekhizdat, 1950, Chapter III). For a given number n of nodes the Gaussian quadrature formula accurately integrates polynomials of up to the $(2n-1)$th degree.

In actual numerical integration there always arises the question of the magnitude of the possible errors. These errors can be estimated by knowing the magnitudes of the high-order derivatives of the integrand. But in practice this method is only rarely applied since the magnitudes of the required derivatives are difficult to estimate and only very rough estimates can be obtained. Instead, we may check the accuracy differently: we compute the approximate value of the integral by two separate quadrature formulas and assume* that the errors in each of these are of the same order as the difference between the results of the computations by the two different quadrature formulas. When this method is used, the amount of work necessary for integration with an accuracy check depends on the total number of points at which the computation of the function is carried out for both quadratures. If Gaussian quadratures of orders n and $n+1$ are used, the total number of points equals $2n+1$.

Naturally, the question arises whether this method is the best. First of all we note that if we wish to carry out the integration by two different quadratures $A(f)$ and $B(f)$, each of which must accurately integrate a polynomial of the $(2n-1)$th degree, then the minimum number of integration nodes needed (for both quadratures together) is exactly $2n+1$. Obviously then we would wish the "more accurate" of these two quadratures to give a correct result for polynomials of possibly higher degree if it had to compute the function at the $2n+1$ points all by itself. The Gaussian quadrature of order $n+1$ does not satisfy this requirement. It accurately integrates polynomials of only up to the $(2n+1)$th degree. It turns out that for the same amount of labor we can in conjunction with a Gaussian quadrature of order n have a quadrature which is accurate for polynomials of up to degree $3n+1$ (for even n) or $3n+2$ (for odd n). Clearly it is impossible to go higher. This point is briefly discussed in section 2.

---

*Such an assumption is based mainly on practical experience. A corresponding theorem holds for certain restricted classes of functions.

2. OPTIMAL QUADRATURE FORMULAS

Let us first note that it suffices to consider the the quadratures on the interval $[-1,+1]$: the integral $\int_a^b f(x)\,dx$ reduces to $\dfrac{b-a}{2}\int_{-1}^{+1} f[x(t)]\,dt$ by the substitution $t = 2\dfrac{x-a}{b-a} - 1$.

Let us designate the *value* of a quadrature A of form (1) by the number $V(A) = n$, and the *accuracy* of the quadrature by an integer $T(A)$ such that polynomials of degree up to $T(A)$ inclusive are accurately integrated by formula (1).

Let $x_1, x_2, \ldots, x_n$ be arbitrary, distinct numbers in $[-1,+1]$. There holds

Theorem 1. *The weights* $V_1, V_2, \ldots, V_n$ *exist and are unique, and for the quadrature*

$$A(f) = \sum_{i=1}^n V_i f(x_i) \qquad (2)$$

ensure an accuracy not less than $n-1$, *i.e.,* $T(A) \geq n-1$.

Proof. By assuming $f_k(x) = x^k$, ($k = 0, 1, 2, \ldots, n-1$), from (2) we get the system of equations

$$\left.\begin{aligned}
V_1 + V_2 + \ldots + V_n &= 2/1, \\
V_1 x_1 + V_2 x_2 + \ldots + V_n x_n &= 0, \\
V_1 x_1^2 + V_2 x_2^2 + \ldots + V_n x_n^2 &= 2/3, \\
\cdots\cdots\cdots\cdots\cdots\cdots \\
V_1 x_1^{n-1} + V_2 x_2^{n-1} + \ldots + V_n x_n^{n-1} &= \begin{cases} \text{for even } n \\ \text{for odd } n \end{cases}
\end{aligned}\right\} \quad (3)$$

The determinant of system (3) is the Vandermonde determinant, whence the assertion of the theorem follows.

Now, let the nodes $x_1, x_2, \ldots, x_n$ be chosen not arbitrarily but as the roots of the Legendre polynomial $L_n(x)$ of order n. Let us designate these by $a_1, a_2, \ldots, a_n$. We can compute the weights $V_1, V_2, \ldots, V_n$ by applying theorem 1. The quadrature so obtained is also a Gaussian quadrature of order n

$$G_n(f) = \sum_{i=1}^n V_i f(a_i). \qquad (4)$$

Theorem 2. *The Gaussian quadrature* G_n *has an accuracy of* $2n-1$.

Proof. By theorem 1 quadrature (4) has an accuracy of not less than $n-1$. It remains to prove that (4) is accurate for polynomials of degrees from n to $2n-1$.

It is well known that the Legendre polynomial $L_n(x)$ of degree n is orthogonal on $[-1,+1]$ to all polynomials of lower degree. Hence it follows that $\int_{-1}^{+1} L_n(x)\,x^k\,dx = 0$ when $k = 0, 1, 2, \ldots, n-1$. Since $L_n(a_i) = 0$, formula (4) gives a zero on the right-hand side for the functions $L_n(x)\,x^k$ ($k = 0, 1, 2, \ldots, n-1$). Consequently, polynomials of the form $x^k L_n(x)$ ($k = 0, 1, 2, \ldots, n-1$) are integrable accurately. But then any polynomial of degree not higher than $2n-1$ is also integrable accurately by formula (4) since any such polynomial can be put in the form

$$P_{n-1}(x) + \sum_{s=0}^{n-1} C_s x^s L_n(x), \qquad (5)$$

where $P_{n-1}(x)$ is a polynomial of degree not higher than $n-1$. Since every term in the summation in (5) is integrable accurately by the Gaussian quadrature, the sum is also integrable accurately and the theorem is proved.

From theorems 1 and 2 we readily obtain

Theorem 3. *The Gaussian quadrature is unique, and when its value is* n *it has an accuracy not less than* $2n-1$.

Proof. Let

$$B(f) = \sum_{i=1}^n W_i f(\beta_i). \qquad (6)$$

be some quadrature of accuracy $T(B) \geq 2n-1$. Let us combine all the points $\{a_i\}$ and $\{\beta_i\}$. This will give $x_1, x_2, \ldots, x_m$, where $m \leq 2n$. By applying theorem 1 we find, uniquely, the weight which en- an accuracy of $m-1 \leq 2n-1$ for the quadrature

$$C(f) = \sum_{i=1}^m R_i f(x_i) \qquad (7)$$

However, the quadrature whose weight is V_i at the points x_s which coincide with the points a_i and is zero at the remaining points already has this property. On the other hand, by setting

$$R_i = \begin{cases} W_s \text{ when } x_i = \beta_s, \\ 0 \text{ when } x_i \neq \beta_s, \end{cases}$$

in (7) we also find that this is a unique quadrature. Hence it follows that quadrature (6) coincides with the Gaussian quadrature and the theorem is proved.

Thus, for a given value, the Gaussian quadrature, and only it, has maximum accuracy.

Now, let A and B be two Gaussian quadratures on $[-1, +1]$. We shall call the set of the two different quadratures the *quadrature pair* (A, B). The accuracy $T(A, B)$ of the quadrature pair will be designated by

$$T(A, B) = \min \{T(A); T(B)\}.$$

In everything that follows we shall consider that $T(A) \le T(B)$ for definiteness.

The value $V(A, B)$ of the quadrature pair (A, B) will be the sum of the number of different nodes involved in both quadratures.

Theorem 4. *For any quadrature pair* (A, B)

$$T(A, B) \leqslant V(A, B) - 2. \tag{8}$$

Proof. Let $C_1, C_2, \ldots, C_{V(A,B)}$ be the set of all the nodes involved in quadratures A and B. Then by theorem 1 there exists a unique quadrature with nodes $\{C_i\}$ and accuracy $T \ge V(A, B) - 1$. But, if $T(A, B) > V(A, B) - 2$, then $T(A) \ge V(A, B) - 1$ and $T(B) \ge V(A, B) - 1$. Hence it follows that quadratures A and B (using one and the same nodes $\{C_i\}$) coincide, i.e., do not form a pair, and the theorem is proved.

Let us now consider the quadrature pair (A, B) whose accuracy is $2n - 1$. From theorem 4 it follows that such a pair has the value $V(A, B) \ge 2n + 1$. Consequently, the most value-economical pair (if such exists!) contains precisely $2n + 1$ nodes. Naturally, they can be distributed in different ways. For example, we could take as quadrature B a Gaussian quadrature of order $2n + 1$ while as A we take the quadrature resulting from theorem 1 if we take $2n$ nodes by rejecting any one of the nodes of quadrature B. However, in such a way we get "nearly similar" quadratures which use almost the same nodes and which, moreover, have similar corresponding weights. Mutual control of the accuracy by such quadratures is unreliable.

Let us require that for an "optimal" pair the number of nodes in the "lower" quadrature is a minimum. Then, quadrature A should coincide with the Gaussian quadrature G_n as a consequence of

theorems 2 and 3. Further, we desire to obtain maximum accuracy from quadrature B within the given value $2n + 1$. We see that such a quadrature is determined uniquely and has an accuracy of $3n + 1$ for even n and of $3n + 2$ for odd n.

Another statement leading to the same result is: given a Gaussian quadrature G_n of order $\hbar$, find a quadrature B different from G_n such that the pair (G_n, B) has the maximum possible accuracy $2n - 1$ and the minimum value $2n + 1$, and in which, moreover, the quadrature B itself has the maximum accuracy possible under these conditions.

Let us now construct such a quadrature B. Let $L_n(x)$ be a Legendre polynomial of degree n. Let us find the polynomial $K_{2n+1}(x) = L_n(x)P_{n+1}(x)$ of degree $2n + 1$ such that it is orthogonal on $[-1, +1]$ to all powers x^k from $k = 0$ to $k = n$. Here the polynomials $L_n(x)$ and $P_{n+1}(x)$ will be normalized so that the coefficients of the terms of highest degree equal 1. Let $M_{n,s}$ denote the sth moment of polynomial $L_n(x)$

$$M_{n,s} = \int_{-1}^{+1} L_n(x) \, x^s dx. \tag{9}$$

Further, let us set

$$P_{n+1}(x) = x^{n+1} + p_n x^n + p_{n-1} x^{n-1} + \ldots + p_0. \tag{10}$$

From the conditions imposed on $K_{2n+1}(x)$ we should have

$$\int_{-1}^{+1} K_{2n+1}(x) \, x^k \, dx = \int_{-1}^{+1} L_n(x) P_{n+1}(x) \, x^k \, dx = 0$$

$$(k = 0, 1, 2, \ldots, n). \tag{11}$$

But $L_n(x)$ is orthogonal on $[-1, +1]$ to all the powers x^r for $r < n$; therefore,

$$\int_{-1}^{+1} L_n(x) \, P_{n+1}(x) \, x^k dx =$$

$$= \int_{-1}^{+1} L_n(x) \, [x^{n+1} + p_n x^n + \ldots + p_0] \, x^k dx =$$

$$= \int_{-1}^{+1} L_n(x)[x^{n+k+1} + p_n x^{n+k} + \ldots + p_{n-k} x^n] \, dx =$$

$$= M_{n, n+k+1} + p_n M_{n, n+k} + \ldots + p_{n-k} M_{n, n}.$$

$$\tag{12}$$

Hence we get a recurrence formula for p_s:

$$
\left.
\begin{aligned}
p_{n+1} &= 1, \\
p_n &= -\frac{M_{n,\,n+1}}{M_{n,\,n}}, \\
p_{n-1} &= -\frac{M_{n,\,n+2}\,p_{n+1} + M_{n,\,n+1}\,p_n}{M_{n,\,n}}, \\
&\cdots\cdots\cdots\cdots\cdots\cdots \\
p_{n-k} &= -\frac{M_{n,\,n+k+1}\,p_{n+1} + \ldots + M_{n,\,n+1}\,p_{n-k+1}}{M_{n,\,n}}, \\
&\cdots\cdots\cdots\cdots\cdots\cdots
\end{aligned}
\right\}
\tag{13}
$$

This also proves the existence of the polynomial $K_{2n+1}(x)$ which we needed.

All the roots of the polynomial $K_{2n+1}(x)$ lie in the interval $[-1, +1]$ and are distributed in increasing order such that the even roots are the roots of the Legendre polynomial $L_n(x)$, while the odd roots are the roots of $P_{n+1}(x)$. I do not know the proof of this in the general case. However, up to $n = 40$ these roots have been computed and are distributed as stated. Therefore, there is nothing to interfere with our taking advantage of what we have said, within the prescribed limits.

Let us denote the roots of $K_{2n+1}(x)$ by $\varkappa_1, \varkappa_2, \ldots, \varkappa_{2n+1}$. By theorem 1 we can find a quadrature

$$
B(f) = \sum_{i=1}^{2n+1} W_i f(\varkappa_i), \tag{14}
$$

having an accuracy not less than $2n$.

Theorem 5. *Quadrature $B(f)$ has an accuracy of $3n+1$ for even n and of $3n+2$ for odd n.*

Proof. Formula (14) is accurate for polynomials of degree $2n$ by hypothesis. It is also precise for the polynomial $K_{2n+1}(x)$ since $\int_{-1}^{+1} K_{2n+1}(x)\,dx = 0$ and the nodes $\varkappa_i$ are the roots of $K_{2n+1}(x)$ and, therefore, $B(K_{2n+1}) = 0 = \int_{-1}^{+1} K_{2n+1}(x)\,dx$.

Furthermore, for polynomials of the form

$$
Q_{2n+1+k}(x) = K_{2n+1}(x)\,x^k \qquad (k = 1, 2, \ldots, n),
$$

having a degree of precisely $2n+1+k$, quadrature B is accurate for the very same reason:

$$
\int_{-1}^{+1} Q_{2n+1+k}(x)\,dx = 0
$$

by virtue of the choice of $K_{2n+1}(x)$, while $B(Q_{2n+1+k}) = 0$ since the nodes $\varkappa_i$ are the roots of $K_{2n+1}(x)$. But, obviously, any polynomial of degree not higher than $3n+1$ has the form

$$
R_{2n}(x) + \sum_{s=0}^{n} C_s x^s K_{2n+1}(x), \tag{15}
$$

where $R_{2n}(x)$ is a polynomial of degree not higher than $2n$. Since all the terms in formula (15) are integrable accurately by quadrature B, the whole sum is integrable accurately. Thus, $T(B) \geq 3n+1$ and the assertion is proved for even n.

Let us note that the polynomial $L_n(x)$ contains powers of x of the same parity as that of n. Hence it follows that all the moments of $L_n(x)$ in $[-1, +1]$ which have opposite parity equal zero. But then from formula (13) it follows that in the polynomial $P_{n+1}(x)$ only those coefficients of the powers of x which have the same parity as $n+1$ are different from zero. Therefore, the product of $P_{n+1}(x)$ and $L_n(x)$, i.e., $K_{2n+1}(x)$, always contains only odd powers of x. Thus,

$$
\int_{-1}^{+1} K_{2n+1}(x)\,x^{2m}\,dx = 0
$$

for any m. Consequently, for odd n quadrature B correctly integrates not only $K_{2n+1}(x)\,x^k$ when $k \leq n$, but also $K_{2n+1}(x)\,x^{n+1}$ and, at the same time, also all polynomials of degree $3n+2$. Thus, the theorem is proved.

Theorem 6. *Among all the quadratures which form a pair with a value of $2n+1$, together with the Gaussian quadrature G_n of order n, only one has an accuracy of $T \geq 3n+1$.*

Proof. Let D be a quadrature with an accuracy of $T \geq 3n+1$ which is different from the quadrature B we have constructed. Let G_n and D form a pair and let $V(G_n, D) \leq 2n+1$. In this case, among the nodes of D not more than $n+1$ are different from the nodes of G_n. Let us combine all the nodes of quadratures B and D. Then there will always be no more than $3n+2$ of them. By theorem 1 there exists only one quadrature with such nodes and with an accuracy not lower then $3n+1$. Since quadrature B meets these requirements, the theorem is proved.

Thus, from theorems 4, 5, and 6 there follows the uniqueness of the quadrature pair (G_n, B) which is optimal in the stated sense.

4

In what follows the quadrature B which has been constructed, when paired with the Gaussian quadrature of order n, will be called an *improved quadrature* of order n.

3. VALUES OF THE REMAINDER TERMS

In comparison with other quadratures, the Gaussian and the improved quadratures usually give a better payoff in accuracy for the same amount of work.

Often this is obtained also when integrating functions whose Taylor series expansions contain an appreciable number of terms which are not integrable exactly. The results of applying Gaussian and improved quadratures to these terms still remain good.

The fact is that the Gaussian quadrature of order n when applied to polynomials of degree $2n$, $2n+2$, etc., gives small relative errors which rapidly fall off with increasing n. Thus, when computing $\int\limits_{-1}^{+1} x^s \, dx$ for $n=10$ and $n=15$, using Gaussian quadratures, the relative errors are:

$n = 10$		$n = 15$	
s	Relative error	s	Relative error
20	$0.307 \cdot 10^{-4}$	30	$0.446 \cdot 10^{-7}$
22	$0.176 \cdot 10^{-3}$	32	$0.368 \cdot 10^{-6}$
24	$0.573 \cdot 10^{-3}$	34	$0.166 \cdot 10^{-5}$
...	....	...	....
...	....	...	....
30	$0.504 \cdot 10^{-2}$	40	$0.325 \cdot 10^{-4}$
32	$0.819 \cdot 10^{-2}$	42	$0.660 \cdot 10^{-4}$

We see, for example, that if we require accuracy in the fourth decimal place then in this sense G_{10} is "accurate" not up to x^{19} but, as follows from theorem 2, up to x^{21}. And for $n = 15$ the "increase in accuracy" in this sense is still greater — instead of $s = 29$ accuracy in the fourth decimal place is preserved up to $s = 43$.

The very same thing holds true also for the improved quadrature. The relative errors of the improved quadrature of $\int\limits_{-1}^{+1} x^s \, dx$ for $n = 10$ and $n = 15$ are:

$n = 10$		$n = 15$	
s	Relative error	s	Relative error
32	$0.726 \cdot 10^{-10}$	48	$0.127 \cdot 10^{-14}$
34	$0.708 \cdot 10^{-9}$	50	$0.170 \cdot 10^{-13}$
36	$0.374 \cdot 10^{-8}$	52	$0.121 \cdot 10^{-12}$
...	....	...	....
...	....	...	....
66	$0.658 \cdot 10^{-4}$	94	$0.337 \cdot 10^{-6}$
68	$0.897 \cdot 10^{-4}$	96	$0.466 \cdot 10^{-6}$

For the same accuracy in the fourth decimal place the improved quadrature with $n = 10$ correctly integrates on $[-1, +1]$ polynomials of degree $s = 69$ instead of the theoretical degree of $s = 31$. For $n = 15$ polynomials of degree up to 96, instead of $s = 41$, are integrable with accuracy greater than the sixth decimal place. Thus, for the improved quadrature this effect is manifest even more sharply than for the Gaussian quadrature.

4. COMPUTATION OF THE NODES AND WEIGHTS OF GAUSSIAN AND IMPROVED QUADRATURES ON AN ELECTRONIC COMPUTER

Formula (13) makes it possible to compute the coefficients of the polynomial $P_{n+1}(x)$ from the moments of the Legendre polynomials $L_n(x)$. To compute the moments

$$M_{n, s} = \int\limits_{-1}^{+1} L_n(x) \, x^s \, dx,$$

or more precisely, to compute their sequential ratios, which alone are used in formula (13), we can use the easily verifiable relation

$$M_{n, n+2s} = M_{n, n+2s-2} \cdot \left(1 + \frac{\left[\frac{n}{2}\right]}{s}\right) \cdot \frac{2\left[\frac{n+1}{2}\right] + 2s - 1}{2n + 2s + 1}. \tag{16}$$

We avoid troublesome interferences when we compute by this formula.

Since modern computers operate with numbers (in the decimal system) of 10-12 or more digits, it is desirable for the supporting tables to have 14-16 digits. Since the literature lacks tables of Gaussian nodes and weights with that many digits (and, in general, since there are no tables for large n), all the computations of the nodes and weights of the Gaussian quadratures had to be carried out anew. The recurrence formulas

$$L_0(x) = 1, \quad L_1(x) = x,$$
$$L_n(x) = \left(2 - \frac{1}{n}\right) x L_{n-1}(x) - \left(1 - \frac{1}{n}\right) L_{n-2}(x) \Bigg\}$$
$$(17)$$

were used to compute the values of the Legendre polynomials.

A special program was constructed for seeking the roots of arbitrary functions on the given interval.

The computations were carried out in the following order. For a given n the roots of $L_n(x)$ in $[-1, +1]$ were computed by the root-seeking program; then the values of the functions $L_n(x)$ were computed by formula (17). Since the roots of $L_n(x)$ are symmetric around zero, it sufficed to find only the positive roots $a_1, a_2, \ldots, a_p$, where $p = \left[\dfrac{n}{2}\right]$. Next, the Gaussian weights were found from the system of equations

$$\left. \begin{aligned} V_1 a_1^2 + V_2 a_2^2 + \ldots + V_p a_p^2 &= \frac{1}{3}, \\ V_1 a_1^4 + V_2 a_2^4 + \ldots + V_p a_p^4 &= \frac{1}{5}, \\ \cdots\cdots\cdots\cdots\cdots\cdots \\ V_1 a_1^{2p} + V_2 a_2^{2p} + \ldots + V_p a_p^{2p} &= \frac{1}{2p+1} \end{aligned} \right\} \quad (18)$$

Here, to find V_k from (18) it is convenient to eliminate the other V_s sequentially; the transformation is necessarily carried out only over the free terms and, after the other V_s have been eliminated, the coefficients of V_k are calculated directly.

After the roots of $L_n(x)$ and the Gaussian weights have been obtained, a control check is carried out: the quantities

$$R_{n,2m} = (2m+1)\,[V_1 a_1^{2m} + V_2 a_2^{2m} + \ldots + V_p a_p^{2m}] - 1$$

are computed, where $m = 1, 2, \ldots, n$. The quantities $R_{n,m}$ should equal zero when $m < n$.

When $m = n$ we get the first nonzero remainder term. This quantity gives the relative accuracy of the nth-order Gaussian quadrature when integrating x^{2n}. The following term, $R_{n,2m}$, has the same significance for the function x^{2n+2}, etc.

The next step consisted of computing the moment relations of $L_n(x)$ and then the coefficients of the polynomial $P_{n+1}(x)$ from formulas (16) and (13). Further, the positive roots $b_1, b_2, \ldots, b_t$ $\left(t = \left[\dfrac{n+1}{2}\right]\right)$ of the polynomial $P_{n+1}(x)$ were computed by the use of the same function-root-seeking program, and as a result all the roots $\{x_i\}$ of the polynomial

$$K_{n+1}(x) = L_n(x) P_{n+1}(x)$$

were determined. After this, the weights W_i ($i = 1, 2, \ldots, n$) of the improved quadrature were found for positive nodes by the same method as was used to obtain the Gaussian weights V_i.

For negative nodes the weights equal the weights at the symmetric positive points. The weight of the zeroth node is obtained from the condition that the sum of all the weights equals 2 for the Gaussian as well as the improved quadrature. The accuracy check is carried out in the same way as for Gaussian quadratures. The relative errors when integrating powers of x beginning with $3n+2$ (or $3n+3$) on $[-1, +1]$ are computed in analogous fashion.

For all these computations it was necessary to retain a large number of stored digits during the intermediate calculations. To preserve 16 correct digits reliably in the answer, it turned out that it was insufficient to retain 21 decimal figures in the intermediate calculations.

To obtain the needed accuracy we had to resort to the development of a library of special programs which allowed us to write a number with a large number of digits into several cells.

The double-precision library routine, originally constructed by A. V. Uskov, allowed us to work with such numbers by writing each one into three memory cells: the order and the sign of the number were entered into the first cell, and the 72 bits of the mantissa into the other two. However, as a result we obtained only 10-12 correct figures for large n.

We had to construct a multiprecision library-routine in which the sign and the order of the number were assigned to the leading cell, while the mantissa of $36k$ bits was written into k successive memory cells following the leading cell. The number k was specifiable. We certainly achieved the needed accuracy by using $k = 6$. It is curious to note that for large n the computation of the values of the Legendre polynomials turned out to be particularly troublesome, more so than the computations of the coefficients of $P_{n+1}(x)$ and of the roots of $L_n(x)$ and $P_{n+1}(x)$.

When integrating polynomials of up to the $(2n-2)$th degree by Gaussian quadratures, the nodes and weights obtained ensure a maximum relative error of not more than 10^{-41}, while when integrating polynomials of up to the $(3n+1)$th [or $(3n+2)$th] degree by improved quadratures, the maximum relative error is not more than 10^{-39}.

5. CHECKING OF THE TABLES

The results of the machine calculations were printed out in decimal and octal forms on tabulograms. The tables were printed from these. The possibility of errors was eliminated by the use of photo-printing. However, it would have been impossible to assume that the supporting tables did not have a single error. But a visual check (computation) was not suitable. Moreover, apart from the errors due to reprinting from the tabulograms, it was not possible to exclude the possibility of machine printing errors. All this necessitated a stringent mechanical check.

The following method was used for this purpose. First of all we note that the machine calculations themselves are easily checked: the relative errors during the integration of all the assumed powers of x never reached 10^{-39}, which excluded errors not only in the last digit but even those 10^5 times smaller. This check was carried out after printing.

The tables were printed in the decimal and octal systems for calculations on the interval $[0, 1]$. The nodes and weights are printed out completely in both systems, i.e., besides the corresponding average node, the weight is printed out twice, and the node which is symmetric with respect to ½ is also printed out in both systems and not obtained by counting from unity.

Furthermore, the decimal and octal systems were printed out independently. Therefore, an agreement of the decimal and octal tables guarantees the absence of error. A perfocheck was made, i.e., for each given n, from the prepared tables cards were punched for the values of the nodes, of the Gaussian weights, and of the weights of the improved quadrature, in both the decimal and the octal systems. The punch cards thus obtained were fed into the machine. The numbers were reworked into the multiprecision form* and the modulus of the difference of the corresponding numbers, one from the decimal and the other from the octal version, was compared with a standard of $1.5 \cdot 2^{-48}$. If the difference exceeded the standard, then the number was ruled out. Corrections to the tables were made by repunching the card with the incorrect number. After this, a check of all the numbers referring to the given n was repeated.

Such a method guaranteed the absence of errors exceeding $1.5 \cdot 2^{-48}$. This sufficed for octal values. However, for decimal numbers this guaranteed absence of error only for the first 14 figures. Therefore, a second check was performed on the machine by punch cards of the entries in the decimal tables: it was checked whether the sum of the symmetric nodes equaled unity (with an accuracy up to 10^{-16}) and whether the symmetric weights of the Gaussian and the improved quadratures coincided.

Finally, as an auxiliary check, the expressions

$$M_{n,k} = (k+1) \sum_{i=1}^{n} V_{n,i}\alpha_i^k - 1, \quad k = 0, 1, 2, \ldots, 2n-1,$$

$$N_{n,k} = (k+1) \sum_{i=1}^{2n+1} W_{n,i}\beta_i^k - 1, \quad k = 0, 1, 2, \ldots, 3n+1,$$

were computed for each n, where $\{\alpha_i^n\}$ are the nodes and V_i the weights of the nth-order Gaussian quadrature, while $\{\beta_i^n\}$ and $\{W_{n,i}\}$ are, respectively, the nodes and weights of the improved quadrature (all for integration on the interval $[0,1]$). The quantities $M_{n,k}$ and $N_{n,k}$ were computed for both the decimal and the octal versions and did not exceed, respectively, $3 \cdot 10^{-16}$ and $5 \cdot 10^{-16}$, which testifies to the absence of errors in the weights for the average node. For $k \neq 0$ the maximum magnitudes of $M_{n,k}$ and $N_{n,k}$ did not exceed 10^{-15} and 10^{-14}, respectively. The main purpose for this final check was to decrease the probability of undetected repeated

_____

*See above.

7

equal errors in the last figures of the decimal tables.

6. CONTENTS OF THE TABLES

The tables present the values of the nodes and weights of Gaussian and improved quadratures for n from 1 to 40.

Since the majority of the calculations was carried out by electronic computers which operate in the binary system, the tables had to be brought into both the decimal and the octal forms. For convenience, the reduction to the decimal and to the octal versions was done in turn: the decimal on the even pages, the corresponding octal form on the odd pages.

All the nodes and weights are given for integration on the interval $[0, 1]$. The numbers in the decimal tables are given in natural form.

In the octal tables, the numbers are given in the normal (floating) form, i.e., the number A is represented in the form $A = M \cdot 2^p$, where $\frac{1}{2} \le M < 1$, and p is an integer (which is negative or equal to zero in all cases except one — that of the Gaussian weight in the unit node when $n = 1$, where $p = 1$). The number M is called the *mantissa* and p the *order* of number A. The numbers M and p are brought to the octal form.

Thus, for example, for the No. 2 Gaussian node corresponding to $n = 7$, the entry

$$6407 \quad 2070 \quad 6034 \quad 7353 \quad -5$$

denotes the number $M \cdot 2^{-5}$, where

$$M = 6 \cdot 2^{-3} + 4 \cdot 2^{-6} + 0 \cdot 2^{-9} + 7 \cdot 2^{-12} + \ldots + 3 \cdot 2^{-48}.$$

In the octal version the magnitudes of n in the headings and the numbering of the nodes are also given in octal form so that, let us say, $n = 35$ in the decimal version corresponds to $n = 43$ in the octal version, while the node with No. 25 in the decimal version corresponds to node No. 31 in the octal.

Pages 12-121 contain the tables of nodes and weights of Gaussian and improved quadratures. The node numbers are given in the first and last columns. Since the nodes of the improved quadratures include the Gaussian nodes, the second column in the tables — the column of nodes — is common. The Gaussian weights corresponding to even-numbered nodes are in the third column from the left in the tables. Finally, the fourth column contains the weights of the improved quadratures.

In the decimal version all the numbers are given to 16 places after the decimal point. The errors in the decimal version nowhere exceed 1 in the last digit. As for the octal version, all the mantissas are given with 16 octal digits for ease of punching. However, here the guaranteed accuracy of the number is 2^{-48}. In numbers with large negative orders the latter octal digits were not subjected to a check (for the order -3 this consisted of the last octal digit, for the order -6, the last two, for -11, the last three, etc.). Generally speaking, these digits are correct since the calculations were carried out with a large number of stored digits; however, random errors may occur among the latter digits which are at the limits of the guaranteed accuracy.

For each n the remainder terms are given in the bottom row of the decimal tables. This is the magnitude of the relative error when on the interval $[-1, +1]$ we integrate x^{2n} by Gaussian quadratures and x^{3n+2} or x^{3n+3} by the improved quadratures.

On pages 124-131 are given the tables of relative errors when integrating the even powers of x on the interval $[-1, +1]$ by Gaussian quadratures. These tables are derived only in the decimal version. Only three significant figures are given in these tables. The column headings are the values of n, the order of the Gaussian quadrature, and in the stub are the values of k, the power of x for which the relative error is derived. The value of the relative error itself when integrating on $[-1, +1]$ by $G_n(x^k)$ is given at the intersection of the column headed by n and the row headed by k. For each n the table entries start only with $k = 2n$.

Pages 132-139 give the analogous tables of errors for the improved quadratures. It should be kept in mind that the index n here relates to the improved quadrature corresponding to the nth-order Gaussian quadrature, i.e., the improved quadrature contains not n but $2n+1$ nodes. Consequently, for each n the table entries start with $k = 3n+2$ for even n and with $k = 3n+3$ for odd n.

The auxiliary tables of errors for Gaussian and improved quadratures, respectively, are given on pages 140-143. These present the maximum even power k for which the Gaussian (or improved) quadrature guarantees a relative error not exceeding $\Delta = 10^{-9}, 10^{-8}, \ldots, 10^{-1}$, when integrating x^k on $[-1, +1]$.

The last two tables also give the order of the integration which guarantees a given relative accuracy for certain functions.

7. REMARKS ON THE USE ON THE TABLES

Suppose we wish to compute $\int_a^b f(x)\,dx$ with an nth-order Gaussian quadrature. We use the tables for the given n. The nodes of the Gaussian quadrature G_n are even nodes and are presented in these tables. Therefore, all the expressions take the form

$$(b - a) \sum_{i=1}^{n} V_{2i}\, f[a + (b - a)\, x_{2i}],$$

where the x_i are the nodes having the number i in the tables, and the V_i are the corresponding weights of the Gaussian quadrature.

For the very same n, the corresponding expression for the improved quadrature will be

$$(b - a) \sum_{i=1}^{2n+1} W\, f[a + (b - a)\, x_i],$$

where W_i are the weights of the improved quadrature of order n.

Let us introduce some numbers into the example. Let $a = 2.1$, $b = 7.3$, $n = 3$. We find the appropriate table on page 12. For the Gaussian quadrature we get the expression

$5{,}2 \times [0{,}11846344 \cdot f\,(2.1 + 5.2 \cdot 0.04691008) +$
$+\ 0{,}23931434 \cdot f\,(2.1 + 5.2 \cdot 0.23076534) +$
$+\ 0{,}28444444 \cdot f\,(2.1 + 5.2 \cdot 0{,}50000000) +$
$+\ 0{,}23931434 \cdot f\,(2.1 + 5.2 \cdot 0.76923466) +$
$\qquad +\ 0.11846344 \cdot f\,(2{,}1 + 5{,}2 \cdot 0{,}95308992)],$

and for the improved quadrature

$5.2 \times [0.02129102 \cdot f\,(2.1 + 5.2 \cdot 0.00795732) +$
$+\ 0{,}05761666 \cdot f\,(2.1 + 5.2 \cdot 0.04691008) +$
$+\ 0.09340040 \cdot f\,(2.1 + 5.2 \cdot 0.12291664) +$
$+\ 0.12052017 \cdot f\,(2.1 + 5.2 \cdot 0.23076534) +$
$+\ 0.13642490 \cdot f\,(2.1 + 5.2 \cdot 0.36018479) +$
$+\ 0.14149371 \cdot f\,(2.1 + 5.2 \cdot 0.50000000) +$
$+\ 0.13642490 \cdot f\,(2.1 + 5.2 \cdot 0.63981521) +$
$+\ 0.12052017 \cdot f\,(2.1 + 5.2 \cdot 0.76923466) +$
$+\ 0.09340040 \cdot f\,(2.1 + 5.2 \cdot 0.87708336) +$
$+\ 0.05761666 \cdot f\,(2.1 + 5.2 \cdot 0.95308992) +$
$+\ 0.02129102 \cdot f\,(2.1 + 5.2 \cdot 0.99204268)]$

(we limit ourselves everywhere to the first eight figures).

When integrating on a machine with a check on the accuracy, we recommend that the computations be carried out in parallel, for the Gaussian as well as the improved quadratures. Such a method results in an economy of labor since the Gaussian quadrature nodes are also the even nodes of the improved quadrature.

A question of practical interest arises: what deviation between the values of the Gaussian and of the improved quadratures can be tolerated if we wish to integrate with a given relative accuracy ϵ? In a precise sense such a question is not a valid one. In practice we can assume that when $n \geq 5$ the deviation between the Gaussian and the improved quadratures exceeds the error in the improved quadrature by at least 100 times. Once again we emphasize that the preceding statement has no mathematical meaning.

TABLES OF NODES AND WEIGHTS OF QUADRATURE FORMULAS

$$n = 1$$

No.	Nodes	Weights of Gaussian quadratures	Weights of improved quadratures	No.
I	0,II27 0I66 5379 2582	0,0000 0000 0000 0000	0,2777 7777 7777 7778	I
2	5000 0000 0000 0000	I,0000 0000 0000 0000	4444 4444 4444 4444	2
3	8872 9833 4620 74I8	0,0000 0000 0000 0000	2777 7777 7777 7778	3
Remainder terms		$G_2 = I,0 \times I0^0$	$K_6 = 0,I6 \times I0^0$	

$$n = 2$$

No.	Nodes	Weights of Gaussian quadratures	Weights of improved quadratures	No.
I	0,0370 8995 0II3 7243	0,0000 0000 0000 0000	0,0989 8989 8989 8990	I
2	2II3 2486 5405 I87I	5000 0000 0000 0000	2454 5454 5454 5455	2
3	5000 0000 0000 0000	0000 0000 0000 0000	3III IIII IIII III0	3
4	7886 75I3 4594 8I29	5000 0000 0000 0000	2454 5454 5454 5455	4
5	9629 I004 9886 2757	0000 0000 0000 0000	0989 8989 8989 8990	5
Remainder terms		$G_4 = 0,444 \times I0^0$	$K_8 = 0,I63 \times I0^{-I}$	

$$n = 3$$

No.	Nodes	Weights of Gaussian quadratures	Weights of improved quadratures	No.
I	0,0I97 5436 5645 9899	0,0000 0000 0000 0000	0,0523 28II 30I3 2336	I
2	II27 0I66 5379 2583	2777 7777 7777 7778	I342 4404 4934 I667	2
3	2828 78I2 5326 5987	0000 0000 0000 0000	2006 9870 7387 98II	3
4	0,5000 0000 0000 0000	0,4444 4444 4444 4444	0,2254 5826 9329 2872	4
5	0,7I7I 2I87 4673 40I3	0,0000 0000 0000 0000	0,2006 9870 7387 98II	5
6	8872 9833 4620 74I7	2777 7777 7777 7778	I342 4404 4934 I667	6
7	9802 4563 4354 0I0I	0000 0000 0000 0000	0523 28II 30I3 2336	7
Remainder terms		$G_6 = 0,I60 \times I0^0$	$K_{12} = 0,I82 \times I0^{-2}$	

n = 1

No.	Nodes Mantissa	Order	Weights of Gaussian quadratures Mantissa	Order	Weights of improved quadratures Mantissa	Order	No.
I	7I55 004I 360I 0667	- 3	0000 0000 0000 0000	0	4343 4343 4343 4343	- I	I
2	4000 0000 0000 0000	0	4000 0000 0000 0000	+ I	7070 7070 7070 7070	- I	2
3	7062 2773 64I7 67II	0	0000 0000 0000 0000	0	4343 4343 4343 4343	- I	3

n = 2

No.	Nodes Mantissa	Order	Weights of Gaussian quadratures Mantissa	Order	Weights of improved quadratures Mantissa	Order	No.
I	4576 564I 5277 5276	- 4	0000 0000 0000 0000	0	6253 5467 2545 7040	- 3	I
2	6606 26I3 5I57 63I6	- 2	4000 0000 0000 0000	0	7665 4I57 5530 3373	- 2	2
3	4000 0000 0000 0000	0	0000 0000 0000 0000	0	4764 4764 4764 4764	- I	3
4	6236 3235 0544 03I4	0	4000 0000 0000 0000	0	7665 4I57 5530 3373	- 2	4
5	7550 0505 7I24 0I24	0	0000 0000 0000 0000	0	6253 5467 2545 7040	- 3	5

n = 3

No.	Nodes Mantissa	Order	Weights of Gaussian quadratures Mantissa	Order	Weights of improved quadratures Mantissa	Order	No.
I	5035 I750 23I6 I530	- 5	0000 0000 0000 0000	0	6545 3000 70I5 50I0	- 4	I
2	7I55 004I 360I 0667	- 3	4343 4343 4343 4343	- I	4227 3505 2653 I463	- 2	2
3	44I5 2546 644I 0045	- I	0000 0000 0000 0000	0	6330 I766 2045 I3I3	- 2	3
4	4000 0000 0000 0000	0	7070 7070 7070 7070	- I	7I55 72I0 2530 3377	- 2	4
5	557I 25I4 4557 3755	0	0000 0000 0000 0000	0	6330 I766 2045 I3I3	- 2	5
6	7062 2773 64I7 67II	0	4343 4343 4343 4343	- I	4227 3505 2653 I463	- 2	6
7	7657 0540 573I 4345	0	0000 0000 0000 0000	0	6545 3000 70I5 50I0	- 4	7

$$n = 4$$

No.	Nodes	Weights of Gaussian quadratures	Weights of improved quadratures	No.
I	0,0II7 I987 463I 2I34	0,0000 0000 0000 0000	0,03I4 8868 6832 7365	I
2	0694 3I84 4202 9737	I739 2742 2568 7269	0850 2680 2667 86I4	2
3	I798 5689 I25I 8450	0000 0000 0000 0000	I333 99I7 0226 I422	3
4	3300 0947 8207 57I9	3260 7257 743I 273I	I634 7459 4800 7258	4
5	0,5000 0000 0000 0000	0,0000 0000 0000 0000	0,I732 2I49 0945 0682	5
6	0,6699 9052 I792 428I	0,3260 7257 743I 273I	0,I634 7459 4800 7258	6
7	820I 43I0 8748 I550	0000 0000 0000 0000	I333 99I7 0226 I422	7
8	9305 68I5 5797 0263	I739 2742 2568 7269	0850 2680 2667 86I4	8
9	9882 80I2 5368 7866	0000 0000 0000 0000	03I4 8868 6832 7365	9
Remainder terms		$G_8 = 0{,}522 \times 10^{-I}$	$K_{I4} = 0{,}755 \times 10^{-4}$	

$$n = 5$$

No.	Nodes	Weights of Gaussian quadratures	Weights of improved quadratures	No.
I	0,0079 573I 9952 5788	0,0000 0000 0000 0000	0,02I2 9I0I 8375 5409	I
2	0469 I007 7030 6680	II84 6344 2528 0945	0576 I665 83II 2367	2
3	I229 I663 67I4 5754	0000 0000 0000 0000	0934 0039 8278 2463	3
4	2307 6534 4947 I585	2393 I433 5249 6833	I205 20I6 96I4 3238	4
5	360I 8479 34I9 I084	0000 0000 0000 0000	I364 2490 0956 2795	5
6	0,5000 0000 0000 0000	0,2844 4444 4444 4444	0,I4I4 9370 8928 7456	6
7	0,6398 I520 6580 89I6	0,0000 0000 0000 0000	0,I364 2490 0956 2795	7
8	7692 3465 5052 84I5	2393 I433 5249 6833	I205 20I6 96I4 3238	8
9	8770 8336 3285 4246	0000 0000 0000 0000	0934 0039 8278 2463	9
I0	9530 8992 2969 3320	II84 6344 2528 0945	0576 I665 83II 2367	I0
II	9920 4268 0047 42I2	0000 0000 0000 0000	02I2 9I0I 8375 5409	II
Remainder terms		$G_{I0} = 0{,}I6I \times 10^{-I}$	$K_{I8} = 0{,}840 \times 10^{-5}$	

n = 4

No.	Nodes Mantissa	Order	Weights of Gaussian quadratures Mantissa	Order	Weights of improved quadratures Mantissa	Order	No.
I	6000 2267 4404 3047	– 6	0000 0000 0000 0000	0	40I7 5II0 0048 7732	– 4	I
2	4343 III0 3023 43I7	– 3	544I 5007 5763 6663	– 2	5342 I2I0 2I36 7556	– 3	2
3	5602 6I47 5I70 6060	– 2	0000 0000 0000 0000	0	42II 47I2 6I38 6500	– 2	3
4	52I7 3400 46I0 54I0	– I	5I57 I374 I006 0446	– I	5I66 2742 263I 2570	– 2	4
5	4000 0000 0000 0000	0	0000 0000 0000 0000	0	5426 037I 3644 5460	– 2	5
6	5270 2I77 5473 5I73	0	5I57 I374 I006 0446	– I	5I66 2742 263I 2570	– 2	6
7	6437 2346 054I 6363	0	0000 0000 0000 0000	0	42II 47I2 6I38 6500	– 2	7
I0	7343 4666 7475 4346	0	544I 5007 5763 6663	– 2	5342 I2I0 2I36 7556	– 3	I0
II	77I7 7755 I033 7347	0	0000 0000 0000 0000	0	40I7 5II0 0048 7732	– 4	II

n = 5

No.	Nodes Mantissa	Order	Weights of Gaussian quadratures Mantissa	Order	Weights of improved quadratures Mantissa	Order	No.
I	4045 7553 I727 I326	– 6	0000 0000 0000 0000	0	5346 5200 3476 I460	– 5	I
2	6002 2307 7263 0545	– 4	745I 6366 0667 0034	– 3	7277 756I 7446 6322	– 4	2
3	7673 5667 555I 3I32	– 3	0000 0000 0000 0000	0	5764 4265 2004 7273	– 3	3
4	7304 6700 II46 22I7	– 2	7520 732I I327 I0I2	– 2	7555 I507 2600 I740	– 3	4
5	5606 5044 I2I0 08I7	– I	0000 0000 0000 0000	0	4273 I370 0775 7I7I	– 2	5
6	4000 0000 0000 0000	0	4482 I263 6II5 2746	– I	44I6 I672 0434 0474	– 2	6
7	5074 5355 7273 7630	0	0000 0000 0000 0000	0	4273 I370 0775 7I7I	– 2	7
I0	6II6 62I7 7546 3334	0	7520 732I I327 I0I2	– 2	7555 I507 2600 I740	– 3	I0
II	70I0 42II 0222 6464	0	0000 0000 0000 0000	0	5764 4265 2004 7273	– 3	II
I2	7477 6663 4024 635I	0	745I 6366 0667 0034	– 3	7277 756I 7446 6322	– 4	I2
I3	7737 3202 2460 5064	0	0000 0000 0000 0000	0	5346 5200 3476 I460	– 5	I3

n = 6

No.	Nodes	Weights of Gaussian quadratures	Weights of improved quadratures	No.
I	0,0056 4839 8693 6606	0,0000 0000 0000 0000	0,0151 9807 7059 9099	I
2	0337 6524 2898 4240	0856 6224 6189 5852	0418 4722 0223 4533	2
3	0893 1332 9567 4860	0000 0000 0000 0000	0686 6030 2317 2235	3
4	1693 9530 6766 8677	1803 8078 6524 0693	0905 3599 7161 5688	4
5	2684 4089 3762 3477	0000 0000 0000 0000	1066 0482 6135 9811	5
6	0,3806 9040 6958 4015	0,2339 5696 7286 3455	0,1168 8543 2058 4972	6
7	5000 0000 0000 0000	0000 0000 0000 0000	1205 3629 0086 7324	7
8	6193 0959 3041 5985	2339 5696 7286 3455	1168 8543 2058 4972	8
9	0,7315 5910 6237 6523	0,0000 0000 0000 0000	0,1066 0482 6135 9811	9
10	8306 0469 3233 1323	1803 8078 6524 0693	0905 3599 7161 5688	10
11	9106 8667 0432 5140	0000 0000 0000 0000	0686 6030 2317 2235	11
12	9662 3475 7101 5760	0856 6224 6189 5852	0418 4722 0223 4533	12
13	9943 5160 1306 3394	0000 0000 0000 0000	0151 9807 7059 9099	13
Remainder terms		$G_{12} = 0{,}480 \times 10^{-2}$	$K_{20} = 0{,}139 \times 10^{-2}$	

n = 7

No.	Nodes	Weights of Gaussian quadratures	Weights of improved quadratures	No.
I	0,0042 7231 4439 5937	0,0000 0000 0000 0000	0,0114 6766 1005 2646	I
2	0254 4604 3828 6207	0647 4248 3084 4348	0315 4604 6314 9893	2
3	0675 6778 8320 1155	0000 0000 0000 0000	0523 9500 5161 1251	3
4	1292 3440 7200 3028	1398 5269 5744 6383	0703 2662 9857 7630	4
5	2069 5638 2266 1544	0000 0000 0000 0000	0845 0236 3319 6340	5
6	0,2970 7742 4311 3014	0,1909 1502 5252 5595	0,0951 7528 9032 3927	6
7	3961 0752 4496 0508	0000 0000 0000 0000	1022 1647 0037 6494	7
8	5000 0000 0000 0000	2089 7959 1836 7347	1047 4107 0542 3639	8
9	6038 9247 5503 9492	0000 0000 0000 0000	1022 1647 0037 6494	9
10	7029 2257 5688 6986	1909 1502 5252 5595	0951 7528 9032 3927	10
II	0,7930 4361 7733 8456	0,0000 0000 0000 0000	0,0845 0236 3319 6340	11
12	8707 6559 2799 6972	1398 5269 5744 6383	0703 2662 9857 7630	12
13	9324 3221 1679 8845	0000 0000 0000 0000	0523 9500 5161 1251	13
14	9745 5395 6171 3793	0647 4248 3084 4348	0315 4604 6314 9893	14
15	9957 2768 5560 4063	0000 0000 0000 0000	0114 6766 1005 2646	15
Remainder terms		$G_{14} = 0{,}139 \times 10^{-2}$	$K_{24} = 0{,}717 \times 10^{-7}$	

16

n = 6

No.	Nodes Mantissa	Order	Weights of Gaussian quadratures Mantissa	Order	Weights of improved quadratures Mantissa	Order	No.
I	562I 3063 65I7 75I5	− 7	0000 0000 0000 0000	0	7620 0532 7674 6470	− 6	I
2	4244 6554 2775 3770	− 4	5366 7660 036I 3337	− 3	5266 3775 5III 6243	− 4	2
3	5556 4750 I3I3 3272	− 3	0000 0000 0000 0000	0	43II 6705 6I76 I470	− 3	3
4	5327 2766 4652 I552	− 2	56I3 2675 5274 5I40	− 2	5626 5357 6565 5604	− 3	4
5	4227 0425 5563 6220 ·	− I	0000 0000 0000 0000	0	6645 I64I 4854 6I46	− 3	5
6	6056 4732 277I I424	− I	737I II52 23I2 5057	− 2	7366 064I I0I0 7603	− 3	6
7	4000 0000 0000 0000	0	0000 0000 0000 0000	0	7555 5672 7763 5276	− 3	7
I0	4750 5422 6403 3I65	0	737I II52 23I2 5057	− 2	7366 064I I0I0 7603	− 3	I0
II	5664 3565 II06 0667	0	0000 0000 0000 0000	0	6645 I64I 4854 6I46	− 3	II
I2	65I2 I202 2625 3445	0	56I3 2675 5274 5I40	− 2	5626 5357 6565 5604	− 3	I2
I3	7222 I302 7646 4450	0	0000 0000 0000 0000	0	43II 6705 6I76 I470	− 3	I3
I4	7565 545I I640 I200	0	5366 7660 036I 3337	− 3	5266 3775 5III 6243	− 4	I4
I5	7750 6723 4605 300I	0	0000 0000 0000 0000	0	7620 0532 7674 6470	− 6	I5

n = 7

No.	Nodes Mantissa	Order	Weights of Gaussian quadratures Mantissa	Order	Weights of improved quadratures Mantissa	Order	No.
I	4277 7305 3I36 6235	− 7	0000 0000 0000 0000	0	5676 I333 I752 5643	− 6	I
2	6407 2070 6034 7353	− 5	4III 3664 770I I552	− 3	4023 3I55 2475 54I6	− 4	2
3	4246 0373 02I4 4230	− 3	0000 0000 0000 0000	0	655I 6045 0644 0246	− 4	3
4	4I05 3006 2034 2742	− 2	4363 26I3 4236 6062	− 2	4400 3550 3643 3506	− 3	4
5	6476 6I37 5542 0I35	− 2	0000 0000 0000 0000	0	5320 7623 I562 I752	− 3	5
6	460I 52I0 I67I 40I2	− I	6067 7472 356I 453I	− 2	6056 5503 0344 0I65	− 3	6
7	6254 7232 7382 I747	− I	0000 0000 0000 0000	0	6425 3336 2766 2346	− 3	7
I0	4000 0000 0000 0000	0	6537 7277 0036 0575	− 2	6550 II74 4I03 4II0	− 3	I0
II	465I 4262 4222 70I4	0	0000 0000 0000 0000	0	6425 3336 2766 2346	− 3	II
I2	5477 I273 7043 I772	0	6067 7472 356I 453I	− 2	6056 5503 0344 0I65	− 3	I2
I3	6260 2350 0447 3750	0	0000 0000 0000 0000	0	5320 7623 I562 I752	− 3	I3
I4	6756 5I76 3370 7207	0	4363 26I3 4236 6062	− 2	4400 3550 3643 3506	− 3	I4
I5	7353 I740 4756 3354	0	0000 0000 0000 0000	0	655I 6045 0644 0246	− 4	I5
I6	7627 6I36 I637 06I0	0	4III 3664 770I I552	− 3	4023 3I55 2475 54I6	− 4	I6
I7	7756 4002 3523 2046	0	0000 0000 0000 0000	0	5676 I333 I752 5643	− 6	I7

17

No.	Nodes	Weights of Gaussian quadratures	Weights of improved quadratures	No.
I	0,0033 1006 2059 1419	0,0000 0000 0000 0000	0,0089 1119 1660 3552	I
2	0198 5507 1751 2319	0506 1426 8145 1881	0247 1969 7501 0697	2
3	0529 3954 6576 2718	0000 0000 0000 0000	0412 4114 9465 6792	3
4	1016 6676 1293 1866	1111 9051 7226 6872	0558 2318 5413 4198	4
5	1638 2296 4527 4207	0000 0000 0000 0000	0681 3155 4627 5861	5
6	0,2372 3379 5041 8355	0,1568 5332 2938 9436	0,0783 2630 3084 0942	6
7	3196 4945 1035 9340	0000 0000 0000 0000	0860 3530 4277 6057	7
8	4082 8267 8752 1751	1813 4189 1689 1810	0907 0001 2534 0173	8
9	0,5000 0000 0000 0000	0,0000 0000 0000 0000	0,0922 2320 2872 3458	9
I0	0,5917 1732 1247 8249	0,1813 4189 1689 1810	0,0907 0001 2534 0173	I0
II	6803 5054 8964 0660	0000 0000 0000 0000	0860 3530 4277 6057	II
I2	7627 6620 4958 1645	1568 5332 2938 9436	0783 2630 3084 0942	I2
I3	0,8361 7703 5472 5793	0,0000 0000 0000 0000	0,0681 3155 4627 5861	I3
I4	8983 3323 8706 8134	1111 9051 7226 6872	0558 2318 5413 4198	I4
I5	9470 6045 3423 7282	0000 0000 0000 0000	0412 4114 9465 6792	I5
I6	9801 4492 8248 7681	0506 1426 8145 1881	0247 1969 7501 0697	I6
I7	9966 8993 7940 8581	0000 0000 0000 0000	0089 1119 1660 3552	I7
Remainder terms		$G_{16} = 0,396 \times 10^{-3}$	$K_{26} = 0,646 \times 10^{-8}$	

No.	Nodes		Weights of Gaussian quadratures		Weights of improved quadratures		No.
	Mantissa	Order	Mantissa	Order	Mantissa	Order	
I	66I6 6640 2252 77I4	-I0	0000 0000 0000 0000	0	4440 0077 I377 4652	- 6	I
2	5052 3432 365I 47I4	- 5	6365 0350 II45 0033	- 4	6250 0366 4253 040I	- 5	2
3	66I5 3443 2465 63I4	- 4	0000 0000 0000 0000	0	52I6 6I72 6064 27I3	- 4	3
4	6403 325I 5554 0574	- 3	7073 3732 46I6 4534	- 3	7II2 3332 I665 43I7	- 4	4
5	5I74 0465 0276 52II	- 2	0000 0000 0000 0000	0	4270 42I6 357I 5744	- 3	5
6	7456 6552 37I5 3664	- 2	50II 7050 2422 5260	- 2	5006 46I2 3424 653I	- 3	6
7	5072 4427 6II4 2I63	- I	0000 0000 0000 0000	0	5403 I507 0426 I7I2	- 3	7
I0	6420 5I55 3030 343I	- I	5633 0660 26I4 6232	- 2	5634 0355 47I2 2245	- 3	I0
II	4000 0000 0000 0000	0	0000 0000 0000 0000	0	57I5 7604 6042 3435	- 3	II
I2	4567 53II 2363 6I63	0	5633 0660 26I4 6232	- 2	5634 0355 47I2 2245	- 3	I2
I3	5342 5564 073I 6706	0	0000 0000 0000 0000	0	5403 I507 0426 I7I2	- 3	I3
I4	6064 2245 30I4 5022	0	50II 7050 2422 5260	- 2	5006 46I2 3424 653I	- 3	I4
I5	6540 7662 5720 2535	0	0000 0000 0000 0000	0	4270 42I6 357I 5744	- 3	I5
I6	7I37 4452 6222 3720	0	7073 3732 46I6 4534	- 3	7II2 3332 I665 43I7	- 4	I6
I7	7447 I2I5 6254 5063	0	0000 0000 0000 0000	0	52I6 6I72 6064 27I3	- 4	I7
20	7656 5307 I302 546I	0	6365 0350 II45 0033	- 4	6250 0366 4253 040I	- 5	20
2I	7762 3422 2778 2520	0	0000 0000 0000 0000	0	4440 0077 I377 4652	- 6	2I

No.	Nodes	Weights of Gaussian quadratures	Weights of improved quadratures	No.
I	0,0026 609I 966I 3299	0,0000 0000 0000 0000	0,007I 5238 782I 9I95	I
2	0I59 I988 0246 I870	0406 37I9 4I80 7872	0I98 I594 7580 I306	2
3	0425 I824 6375 I6II	0000 0000 0000 0000	0332 5907 7970 I37I	3
4	08I9 8444 6336 682I	0903 2408 0347 4287	0453 9534 0844 3632	4
5	I327 566I 7408 033I	0000 0000 0000 0000	0558 9456 7342 209I	5
6	0,I933 I428 3649 7048	0,I303 0534 820I 4677	0,0650 0070 3427 6706	6
7	2622 6876 0443 770I	0000 0000 0000 0000	0726 I979 4I92 I88I	7
8	3378 7328 8298 0955	I56I 7353 8520 00I4	0782 0676 3894 24I9	8
9	4I78 882I 8I92 5066	0000 0000 0000 0000	08I4 3I4I 3720 0575	9
I0	0,5000 0000 0000 0000	0,I65I I967 7500 6299	0,0824 4800 64I4 I747	I0
II	0,582I II78 I807 4934	0,0000 0000 0000 0000	0,08I4 3I4I 3720 0575	II
I2	662I 267I I70I 9045	I56I 7353 8520 00I4	0782 0676 3894 24I9	I2
I3	7377 3I23 9556 2299	0000 0000 0000 0000	0726 I979 4I92 I88I	I3
I4	8066 857I 6350 2952	I303 0534 820I 4677	0650 0070 3427 6706	I4
I5	0,8672 4338 259I 9669	0,0000 0000 0000 0000	0,0558 9456 7342 209I	I5
I6	9I80 I555 3663 3I79	0903 2408 0347 4287	0453 9534 0844 3632	I6
I7	9574 8I75 3624 8389	0000 0000 0000 0000	0332 5907 7970 I37I	I7
I8	9840 80II 9753 8I30	0406 37I9 4I80 7872	0I98 I594 7580 I306	I8
I9	9973 3908 0338 670I	0000 0000 0000 0000	007I 5238 782I 9I95	I9
Remainder terms		$G_{18} = 0,III \times I0^{-3}$	$K_{30} = 0,739 \times I0^{-3}$	

20

n = II

No.	Nodes		Weights of Gaussian quadratures		Weights of improved quadratures		No.
	Mantissa	Order	Mantissa	Order	Mantissa	Order	
I	5346 I322 7303 270I	−I0	0000 0000 0000 0000	0	7245 7223 7I04 4056	− 7	I
2	4046 5I50 5006 7574	− 5	5I47 I457 600I 7567	− 4	5045 24I5 6627 4033	− 5	2
3	5342 3634 6652 I046	− 4	0000 0000 0000 0000	0	4203 5253 6055 2I43	− 4	3
4	5I76 3566 0745 7654	− 3	56I7 5724 662I 0537	− 3	5637 0I67 0I2I I666	− 4	4
5	4I77 053I 6206 0536	− 2	0000 0000 0000 0000	0	7II7 0663 5302 2II0	− 4	5
6	6I37 2055 76I0 0047	− 2	4I26 7303 7I0I I650	− 2	4I2I 7426 5670 I276	− 3	6
7	4I44 4027 2233 6245	− I	0000 0000 0000 0000	0	45I3 4657 62I4 40I2	− 3	7
I0	58I7 6672 2I55 6526	− I	4776 5764 6034 3250	− 2	5002 5336 I240 I634	− 3	I0
II	6537 256I 6320 5202	− I	0000 0000 0000 0000	0	5I54 2603 2545 5607	− 3	II
I2	4000 0000 0000 0000	0	522I 244I 7662 5I03	− 2	52I5 5200 0625 5256	− 3	I2
I3	4520 2507 0627 5276	0	0000 0000 0000 0000	0	5I54 2603 2545 5607	− 3	I3
I4	5230 0442 67II 0524	0	4776 5764 6034 3250	− 2	5002 5336 I240 I634	− 3	I4
I5	57I5 5764 2662 0655	0	0000 0000 0000 0000	0	45I3 4657 62I4 40I2	− 3	I5
I6	6350 I364 4035 7766	0	4I26 7303 7I0I I650	− 2	4I2I 7426 5670 I276	− 3	I6
I7	6740 I65I 4336 3650	0	0000 0000 0000 0000	0	7II7 0663 5302 2II0	− 4	I7
20	7260 I42I I703 20I2	0	56I7 5724 662I 0537	− 3	5637 0I67 0I2I I666	− 4	20
2I	752I 6606 I445 2735	0	0000 0000 0000 0000	0	4203 5253 6055 2I43	− 4	2I
22	7676 6254 5657 6204	0	5I47 I457 600I 7567	− 4	5045 24I5 6627 4033	− 5	22
23	7765 0635 I32I I7I2	0	0000 0000 0000 0000	0	7245 7223 7I04 4056	− 7	23

No.	Nodes	Weights of Gaussian quadratures	Weights of improved quadratures	No.
I	0,002I 7I4I 8487 0960	0,0000 0000 0000 0000	0,0058 473I 9433 6859	I
2	0I30 4673 574I 4I4I	0333 3567 2I54 344I	0I62 7908 II53 9824	2
3	0349 2I25 4822 I459	0000 0000 0000 0000	0273 7794 8287 I760	3
4	0674 683I 6655 5077	0747 2567 4575 2903	0375 I983 7405 4600	4
5	I095 9II3 6706 79I6	0000 0000 0000 0000	0465 6272 729I 8488	5
6	0,I602 952I 5850 4878	0,I095 43I8 I257 99I0	0,0546 9357 940I I488	6
7	2I86 2I43 2665 6977	0000 0000 0000 0000	06I7 4598 8I3I 0329	7
8	2833 0230 2935 3764	I346 3335 9654 9982	0673 5460 8655 7367	8
9	3528 0356 8649 2699	0000 0000 0000 0000	07I3 8796 9288 5300	9
I0	4255 6283 0509 I844	I477 62II 2357 3764	0738 6955 2450 6692	I0
II	0,5000 0000 0000 0000	0,0000 0000 0000 0000	0,0747 2277 700I 4585	II
I2	0,5744 37I6 9490 8I56	0,I477 62II 2357 3764	0,0738 6955 2450 6692	I2
I3	647I 9643 I350 730I	0000 0000 0000 0000	07I3 8796 9288 5300	I3
I4	7I66 9769 7064 6236	I346 3335 9654 9982	0673 5460 8655 7367	I4
I5	78I3 7856 7334 3023	0000 0000 0000 0000	06I7 4598 8I3I 0329	I5
I6	8397 0478 4I49 5I22	I095 43I8 I257 99I0	0546 9357 940I I488	I6
I7	0,8904 0886 3293 2084	0000 0000 0000 0000	0,0465 6272 729I 8483	I7
I8	9325 3I68 3344 4923	0747 2567 4575 2903	0375 I983 7405 4600	I8
I9	9650 7874 5677 854I	0000 0000 0000 0000	0273 7794 8287 I760	I9
20	9869 5326 4258 5859	0333 3567 2I54 344I	0I62 7908 II53 9824	20
2I	9973 2858 I5I2 9040	0000 0000 0000 0000	0058 473I 9433 6859	2I
Remainder terms		$G_{20} = 0{,}307 \times 10^{-4}$	$K_{32} = 0{,}726 \times 10^{-10}$	

n = 12

No.	Nodes Mantissa	Order	Weights of Gaussian quadratures Mantissa	Order	Weights of improved quadratures Mantissa	Order	No.
I	4344 7133 3065 2154	-10	0000 0000 0000 0000	0	5771 5336 6745 2101	- 7	I
2	6534 0771 6520 6450	- 6	4210 5374 2661 3340	- 4	4125 5665 1124 5002	- 5	2
3	4360 4626 6477 5616	- 4	0000 0000 0000 0000	0	7004 3670 0414 6315	- 5	3
4	4242 6324 1306 4641	- 3	4620 4706 2074 3577	- 3	4632 7146 5233 4427	- 4	4
5	7007 0521 3013 1163	- 3	0000 0000 0000 0000	0	5753 4216 7356 2634	- 4	5
6	5102 2155 6562 0746	- 2	7005 4054 7200 2741	- 3	7000 3137 7324 2014	- 4	6
7	6775 7113 7740 1755	- 2	0000 0000 0000 0000	0	7716 4534 3656 4556	- 4	7
10	4420 6377 6677 5124	- 1	4235 6523 6452 6337	- 2	4237 0466 4264 5106	- 3	10
11	5512 1253 2650 2573	- 1	0000 0000 0000 0000	0	4443 1733 0266 0160	- 3	11
12	6636 1537 0355 6073	- 1	4564 7363 4012 3267	- 2	4564 4353 3771 0365	- 3	12
13	4000 0000 0000 0000	0	0000 0000 0000 0000	0	4620 4101 2676 3210	- 3	13
14	4460 7120 3611 0742	0	4564 7363 4012 3267	- 2	4564 4353 3771 0365	- 3	14
15	5132 7252 2453 6502	0	0000 0000 0000 0000	0	4443 1733 0266 0160	- 3	15
16	5567 4600 0440 1325	0	4235 6523 6452 6337	- 2	4237 0466 4264 5106	- 3	16
17	6200 4155 0007 7404	0	0000 0000 0000 0000	0	7716 4534 3656 4556	- 4	17
20	6557 3344 4243 3606	0	7005 4054 7200 2741	- 3	7000 3137 7324 2014	- 4	20
21	7077 0725 6476 4661	0	0000 0000 0000 0000	0	5753 4216 7356 2634	- 4	21
22	7353 5145 3647 1313	0	4620 4706 2074 3577	- 3	4632 7146 5233 4427	- 4	22
23	7560 7546 4454 0107	0	0000 0000 0000 0000	0	7004 3670 0414 6315	- 5	23
24	7712 4370 0612 5713	0	4210 5374 2661 3340	- 4	4125 5665 1124 5002	- 5	24
25	7767 0661 5111 6253	0	0000 0000 0000 0000	0	5771 5336 6745 2101	- 7	25

No.	Nodes	Weights of Gaussian quadratures	Weights of improved quadratures	No.
1	0,0018 1519 3055 2287	0,0000 0000 0000 0000	0,0048 8272 0522 9804	1
2	0108 8567 0926 9715	0278 3428 3558 0868	0135 7827 7341 0521	2
3	0291 6144 5710 9660	0000 0000 0000 0000	0229 1468 9282 2132	3
4	0564 6870 0115 9524	0627 9018 4732 4523	0315 4871 2375 1875	4
5	0919 7127 1671 8895	0000 0000 0000 0000	0393 3228 5966 1137	5
6	0,1349 2399 7212 9753	0,0931 4510 5463 8671	0,0464 7654 9298 4504	6
7	1847 0023 9919 0175	0000 0000 0000 0000	0529 3603 7240 6947	7
8	2404 5193 5396 5941	1165 9688 2295 9952	0583 6975 1230 5236	8
9	3010 2792 9523 8112	0000 0000 0000 0000	0625 7939 9550 1598	9
10	3652 2842 2023 8275	1314 0227 2255 1233	0656 4034 2114 9028	10
11	0,4319 4349 9600 3191	0,0000 0000 0000 0000	0,0675 9678 6399 9423	11
12	5000 0000 0000 0000	1364 6254 3388 9503	0682 8889 7355 5592	12
13	5680 5650 0399 6809	0000 0000 0000 0000	0675 9678 6399 9423	13
14	0,6347 7157 7976 1725	0,1314 0227 2255 1233	0,0656 4034 2114 9028	14
15	6989 7207 0476 1888	0000 0000 0000 0000	0625 7939 9550 1598	15
16	7595 4806 4603 4059	1165 9688 2295 9952	0583 6975 1230 5236	16
17	8152 9976 0080 9825	0000 0000 0000 0000	0529 3603 7240 6947	17
18	8650 7600 2787 0247	0931 4510 5463 8671	0464 7654 9298 4504	18
19	0,9080 2872 8328 1105	0,0000 0000 0000 0000	0,0393 3228 5966 1137	19
20	9485 3129 9884 0476	0627 9018 4732 4523	0315 4871 2375 1875	20
21	9708 3855 4289 0340	0000 0000 0000 0000	0229 1468 9282 2132	21
22	9891 1432 9073 0285	0278 3428 3558 0868	0135 7827 7341 0521	22
23	9981 8480 6944 7713	0000 0000 0000 0000	0048 8272 0522 9804	23
Remainder terms		$G_{22} = 0{,}843 \times 10^{-5}$	$K_{36} = 0{,}840 \times 10^{-11}$	

No.	Nodes		Weights of Gaussian quadratures		Weights of improved quadratures		No.
	Mantissa	Order	Mantissa	Order	Mantissa	Order	
I	7336 5705 4734 I655	-II	0000 0000 0000 0000	0	4777 7472 366I 4505	- 7	I
2	5445 4720 I20I 3II2	- 6	7I00 227I I455 67II	- 5	6747 3554 2I66 4736	- 6	2
3	7356 I773 7503 7347	- 5	0000 0000 0000 0000	0	5673 3626 I04I 2534	- 5	3
4	7I64 567I 2076 7053	- 4	40II 4043 7367 6I06	- 3	4023 4470 774I 702I	- 4	4
5	5705 5557 I000 2264	- 3	0000 0000 0000 0000	0	502I 5344 0746 60I3	- 4	5
6	4242 4604 I336 3273	- 2	5754 I334 3337 76I4	- 3	5745 706I 5504 43I7	- 4	6
7	5722 I0I7 2200 7750	- 2	0000 0000 0000 0000	0	66I5 I565 2272 4I57	- 4	7
I0	7543 44I0 I662 346I	- 2	7354 5I30 5043 3402	- 3	736I 2436 6234 2646	- 4	I0
II	4642 0I25 I4I7 4607	- I	0000 0000 0000 0000	0	4002 4640 6474 2440	- 3	II
I2	5657 7470 2005 I364	- I	4I50 7I2I I575 I3I0	- 2	4I46 7I6I 4527 2076	- 3	I2
I3	6722 3662 6220 0636	- I	0000 0000 0000 0000	0	4247 0057 0563 4067	- 3	I3
I4	4000 0000 0000 0000	0	4273 6826 2077 0046	- 2	4275 54I4 5I5I I603	- 3	I4
I5	4426 6046 4667 7460	0	0000 0000 0000 0000	0	4247 0057 0563 4067	- 3	I5
I6	5050 0I43 6775 3205	0	4I50 7I2I I575 I3I0	- 2	4I46 7I6I 4527 2076	- 3	I6
I7	5456 7725 3I70 I474	0	0000 0000 0000 0000	0	4002 4640 6474 2446	- 3	I7
20	6047 0675 7423 3063	0	7354 5I30 5043 3402	- 3	736I 2436 6234 2646	- 4	20
2I	64I3 3574 I337 6005	0	0000 0000 0000 0000	0	66I5 I565 2272 4I57	- 4	2I
22	6727 2636 75I0 3I2I	0	5754 I334 3337 76I4	- 3	5745 706I 5504 43I7	- 4	22
23	7207 2222 0677 755I	0	0000 0000 0000 0000	0	502I 5344 0746 60I3	- 4	23
24	7430 5504 3274 0435	0	40II 4043 7367 6I06	- 3	4023 4470 774I 702I	- 4	24
25	76I0 4340 I005 70I0	0	0000 0000 0000 0000	0	5673 3626 I04I 2534	- 5	25
26	7723 3230 5765 7646	0	7I00 227I I455 67II	- 5	6747 3554 2I66 4736	- 6	26
27	7770 44I2 0723 0436	0	0000 0000 0000 0000	0	4777 7472 366I 4505	- 7	27

No.	Nodes	Weights of Gaussian quadratures	Weights of improved quadratures	No.
I	0,00I5 3303 8735 2023	0,0000 0000 0000 0000	0,004I 2885 57I6 5842	I
2	0092 I968 2876 6404	0235 8766 8I93 2559	0II5 I804 20I9 49II	2
3	0247 3II0 2028 4394	0000 0000 0000 0000	0I94 576I 5234 6497	3
4	0479 4I37 I8I4 7626	0534 6966 2997 6592	0268 4850 8803 878I	4
5	0782 2093 79I9 4234	0000 0000 0000 0000	0336 2545 3525 4200	5
6	0,II50 4866 2902 8477	0,0800 39I6 427I 673I	0,0399 60I3 7666 8009	6
7	I579 7005 2264 972I	0000 0000 0000 0000	0457 7473 4I47 5246	7
8	2063 4I02 2856 69I3	I0I5 887I 336I 5330	0508 2486 6I39 530I	8
9	2593 3027 4760 92I5	0000 0000 0000 0000	0550 II30 2488 8220	9
I0	3I60 8425 0500 9099	II67 4626 8269 I774	0583 5602 6750 8784	I0
II	0,3757 47I2 5839 7654	0,0000 0000 0000 0000	0,0608 I3I5 I76I 9742	II
I2	4373 8329 5744 2655	I245 7352 2906 70I4	0622 9208 2268 0780	I2
I3	5000 0000 0000 0000	0000 0000 0000 0000	0627 7844 6952 7372	I3
I4	5626 I670 4255 7345	I245 7352 2906 70I4	0622 9208 2268 0780	I4
I5	6242 5287 4I60 2346	0000 0000 0000 0000	0608 I3I5 I76I 9742	I5
I6	0,6839 I574 9499 090I	0,II67 4626 8269 I774	0,0588 5602 6750 8784	I6
I7	7406 6972 5239 0785	0000 0000 0000 0000	0550 II30 2488 8220	I7
I8	7936 5897 7I43 3087	I0I5 887I 336I 5330	0508 2486 6I39 530I	I8
I9	8420 2994 7735 0279	0000 0000 0000 0000	0457 7473 4I47 5246	I9
20	8849 5I33 7097 I523	0800 39I6 427I 673I	0399 60I3 7666 8009	20
2I	0,92I7 7906 2080 5766	0,0000 0000 0000 0000	0,0336 2545 3525 4200	2I
22	9520 5862 8I85 2374	0534 6966 2997 6592	0268 4850 8803 878I	22
23	9752 6889 797I 5606	0000 0000 0000 0000	0I94 576I 5234 6497	23
24	9907 803I 7I23 3596	0235 8766 8I93 2559	0II5 I804 20I9 49II	24
25	9984 6696 I264 7977	0000 0000 0000 0000	004I 2885 57I6 5842	25
Remainder terms		$G_{24} = 0,229 \times 10^{-5}$	$K_{38} = 0,867 \times 10^{-I2}$	

26

$$n = 14$$

No.	Nodes		Weights of Gaussian quadratures		Weights of improved quadratures		No.
	Mantissa	Order	Mantissa	Order	Mantissa	Order	
I	6217 0076 3545 7461	-II	0000 0000 0000.0000	0	4164 5532 1056 2631	- 7	I
2	4560 7047 0675 0576	- 6	6023 5354 7365 1554	- 5	5713 3053 3446 4447	- 6	2
3	6251 4341 2317 7572	- 5	0000 0000 0000 0000	0	4766 2623 5056 4476	- 5	3
4	6105 7054 0021 6723	- 4	6660 1401 2747 1414	- 4	6677 0547 3204 4756	- 5	4
5	5003 1114 4434 7632	- 3	0000 0000 0000 0000	0	4233 5327 7464 3415	- 4	5
6	7271 7242 1116 7031	- 3	5076 5622 6171 0554	- 3	5072 6475 6154 1326	- 4	6
7	5034 1346 6016 3340	- 2	0000 0000 0000 0000	0	5667 7111 5045 4857	- 4	7
I0	6464 5417 5077 4200	- 2	6400 5437 1530 6533	- 3	6402 6674 0744 4447	- 4	I0
II	4114 3360 0486 5764	- I	0000 0000 0000 0000	0	7025 1610 0431 4400	- 4	II
I2	5035 2713 3725 6471	- I	7361 4252 7022 5525	- 3	7360 3272 5153 3136	- 4	I2
I3	6006 0755 3041 4167	- I	0000 0000 0000 0000	0	7621 3466 0773 7764	- 4	I3
I4	6777 0264 0341 1636	- I	7762 0147 1552 3601	- 3	7762 2773 4044 1030	- 4	I4
I5	4000 0000 0000 0000	0	0000 0000 0000 0000	0	4011 0774 4107 1006	- 3	I5
I6	4400 3645 7617 3060	0	7762 0147 1552 3601	- 3	7762 2773 4044 1030	- 4	I6
I7	4774 7411 2357 1704	0	0000 0000 0000 0000	0	7621 3466 0773 7764	- 4	I7
20	5361 2432 2025 0543	0	7361 4252 7022 5525	- 3	7360 3272 5153 3136	- 4	20
2I	5731 6207 7560 5005	0	0000 0000 0000 0000	0	7025 1610 0431 4400	- 4	2I
22	6262 6474 0560 0737	0	6400 5437 1530 6533	- 3	6402 6674 0744 4447	- 4	22
23	6570 7506 2374 3107	0	0000 0000 0000 0000	0	5667 7III 5045 4857	- 4	23
24	7050 6053 5666 1074	0	5076 5622 6171 0554	- 3	5072 6475 6154 1326	- 4	24
25	7277 4666 3334 3014	0	0000 0000 0000 0000	0	4233 5327 7464 3415	- 4	25
26	7473 5035 1776 7042	0	6660 1401 2747 1414	- 4	6677 0547 3204 4756	- 5	26
27	7632 5470 7531 4004	0	0000 0000 0000 0000	0	4766 2623 5056 4476	- 5	27
30	7732 1707 3071 0272	0	6023 5354 7365 1554	- 5	5713 3053 3446 4447	- 6	30
3I	7771 5607 7014 2320	0	0000 0000 0000 0000	0	4164 5532 1056 2631	- 7	3I

$$n = 13$$

No.	Nodes	Weights of Gaussian quadratures	Weights of improved quadratures	No.
I	0,00I3 I69I I502 5875	0,0000 0000 0000 0000	0,0035 4392 3I75 6243	I
2	0079 0847 2640 7059	0202 4200 2382 6579	0098 7687 3I9I 3530	2
3	02I2 2376 5806 9594	0000 0000 0000 0000	0I67 2I79 4994 7762	3
4	04I2 0080 0388 5II0	0460 6074 99I8 8642	023I 3950 8986 9I54	4
5	0673 334I 9866 8278	0000 0000 0000 0000	0290 5760 52II 5573	5
6	0,0992 I095 4633 3450	0,0694 3675 5I09 8936	0,0346 5I8I 6623 8907	6
7	I365 2557 5339 6840	0000 0000 0000 0000	0899 0298 I084 738I	7
8	I788 2538 0279 8299	0890 7299 0380 9729	0445 8422 0938 7698	8
9	2254 6002 I02I 73I6	0000 0000 0000 0000	0485 7086 7438 0393	9
I0	2757 5362 448I 7766	I039 0802 3768 4443	05I9 I530 0584 5200	I0
II	0,3290 8876 8489 0968	0,0000 0000 0000 0000	0,0546 33I7 5547 6425	II
I2	3847 7084 2022 4326	II3I 4I59 0I3I 4486	0566 05I2 9585 7649	I2
I3	4420 I445 5I27 5332	0000 0000 0000 0000	0577 4439 9545 6432	I3
I4	0.5000 0000 0000 0000	0,II62 7577 66I5 4370	0.058I 0480 6I8I 5304	I4
I5	0,5579 8554 4872 4668	0,0000 0000 0000 0000	0,0577 4439 9545 6432	I5
I6	6I52 29I5 7977 5677	II3I 4I59 0I3I 4486	0566 05I2 9585 7649	I6
I7	6709 I628 I5I0 9032	0000 0000 0000 0000	0546 33I7 5547 6425	I7
I8	0,7242 4637 55I8 2234	0,I039 0802 3768 4443	0,05I9 I530 0584 5200	I8
I9	7745 3997 8978 2684	0000 0000 0000 0000	0485 7086 7438 0393	I9
20	82II 7466 9720 I70I	0890 7299 0380 9729	0445 8422 0938 7698	20
2I	8634 7442 4660 3I60	0000 0000 0000 0000	0899 0298 I084 738I	2I
22	9007 8904 5366 6550	0694 3675 5I09 8936	0346 5I8I 6623 8907	22
23	0,9326 6658 0I33 I722	0,0000 0000 0000 0000	0,0290 5760 52II 5573	23
24	9587 99I9 96II 4890	0460 6074 99I8 8642	023I 3950 8986 9I54	24
25	9787 7623 4I93 0406	0000 0000 0000 0000	0I67 2I79 4994 7762	25
26	9920 9I52 7359 294I	0202 4200 2382 6579	0098 7687 3I9I 3530	26
27	9986 8308 8497 4I25	0000 0000 0000 0000	0035 4392 3I75 6243	27
Remainder terms		$G_{26} = 0{,}620 \times 10^{-6}$	$K_{42} = 0{,}I0I \times 10^{-12}$	

No.	Nodes Mantissa	Order	Weights of Gaussian quadratures Mantissa	Order	Weights of improved quadratures Mantissa	Order	No.
I	5811 6067 5273 0643	-11	0000 0000 0000 0000	0	7204 0452 1070 5465	-10	I
2	4031 1211 6551 4743	- 6	5135 1216 2172 3165	- 5	5035 1233 6537 6613	- 6	2
3	5335 6566 4023 7112	- 5	0000 0000 0000 0000	0	4217 6045 2504 0537	- 5	3
4	5214 1053 5074 6610	- 4	5712 5062 3207 6772	- 4	5730 7421 2260 2200	- 5	4
5	4236 3032 5040 6001	- 3	0000 0000 0000 0000	0	7340 5067 0052 6720	- 5	5
6	6262 7434 7307 0365	- 3	4343 2833 4053 1346	- 3	4336 7420 1445 4171	- 4	6
7	4274 6534 2111 5451	- 2	0000 0000 0000 0000	0	5067 0516 7305 4606	- 4	7
10	5561 6774 6117 7071	- 2	5546 5746 3122 2506	- 3	5551 6761 5331 4057	- 4	10
11	7155 7375 7066 0017	- 2	0000 0000 0000 0000	0	6157 1076 7457 2247	- 4	11
12	4822 7624 1734 7056	- 1	6514 6672 6760 2411	- 3	6512 2443 3054 1247	- 4	12
13	5207 6652 7351 6066	- 1	0000 0000 0000 0000	0	6774 3411 3100 7535	- 4	13
14	6120 0257 0836 0647	- 1	7173 3307 1267 2300	- 3	7175 5307 6133 2415	- 4	14
15	7044 7667 7644 1627	- 1	0000 0000 0000 0000	0	7310 2544 1623 3013	- 4	15
16	4000 0000 0000 0000	0	7342 0776 4360 5334	- 3	7337 7516 1112 2067	- 4	16
17	4355 4044 0055 7064	0	0000 0000 0000 0000	0	7310 2544 1623 3013	- 4	17
20	4727 7650 3620 7454	0	7173 3307 1267 2300	- 3	7175 5307 6133 2415	- 4	20
21	5274 0452 4213 0744	0	0000 0000 0000 0000	0	6774 3411 3100 7535	- 4	21
22	5626 4065 7021 4350	0	6514 6672 6760 2411	- 3	6512 2443 3054 1247	- 4	22
23	6144 4100 4162 3774	0	0000 0000 0000 0000	0	6157 1076 7457 2247	- 4	23
24	6443 4200 6354 0161	0	5546 5746 3122 2506	- 3	5551 6761 5331 4057	- 4	24
25	6720 6250 7355 4465	0	0000 0000 0000 0000	0	5067 0516 7305 4606	- 4	25
26	7151 5034 3047 0741	0	4343 2833 4053 1346	- 3	4336 7420 1445 4171	- 4	26
27	7354 1474 5273 7177	0	0000 0000 0000 0000	0	7340 5067 0052 6720	- 5	27
30	7527 1735 2134 1447	0	5712 5062 3207 6772	- 4	5730 7421 2260 2200	- 5	30
31	7651 0424 2277 3015	0	0000 0000 0000 0000	0	4217 6045 2504 0537	- 5	31
32	7737 4665 6612 2630	0	5135 1216 2172 3165	- 5	5035 1233 6537 6613	- 6	32
33	7772 4661 7102 5047	0	0000 0000 0000 0000	0	7204 0452 1070 5465	-10	33

$$n = 14$$

No.	Nodes	Weights of Gaussian quadratures	Weights of improved quadratures	No.
I	0,00II 3970 3I2I 7284	0,0000 0000 0000 0000	0,0030 6977 9343 I89I	I
2	0068 5809 565I 5938	0I75 5973 0I65 8759	0085 7422 9454 9678	2
3	0I84 2083 0860 5734	0000 0000 0000 0000	0I45 2435 0630 7543	3
4	0357 8255 8I68 2I32	0400 7904 3579 880I	020I 2529 7436 3443	4
5	0585 4266 8373 97I5	0000 0000 0000 0000	0253 4577 I630 2327	5
6	0,0863 9934 2465 II75	0,0607 5928 5343 95I6	0,0303 3356 2983 37II	6
7	II9I 2I62 37I8 8972	0000 0000 0000 0000	0350 5I48 950I 3735	7
8	I563 5854 7594 I573	0786 0I58 3579 0968	0393 2789 8624 8I08	8
9	I976 058I 7029 5892	0000 0000 0000 0000	0430 9I88 I434 729I	9
I0	2423 7568 I820 9230	0927 69I9 8738 9689	0463 684I 5008 9268	I0
II	0,290I 7205 II78 5I05	0,0000 0000 0000 0000	0,049I 3246 3236 05I9	II
I2	3404 438I 5536 055I	I025 9923 I860 6478	05I3 083I 3660 6999	I2
I3	3925 8204 0733 2575	0000 0000 0000 0000	0528 658I 9920 8822	I3
I4	4459 7252 5646 3282	I076 3I92 673I 5789	0538 I32I 0557 0593	I4
I5	0,5000 0000 0000 0000	0,0000 0000 0000 0000	0,054I 3503 3258 2I48	I5
I6	0,5540 2747 4353 67I8	0,I076 3I92 673I 5789	0,0538 I32I 0557 0593	I6
I7	6074 I795 9266 7422	0000 0000 0000 0000	0528 658I 9920 8822	I7
I8	6595 56I8 4463 9449	I025 9923 I860 6478	05I3 083I 3660 6999	I8
I9	7098 2794 882I 4895	0000 0000 0000 0000	049I 3246 3236 05I9	I9
20	0,7576 243I 8I79 0770	0,0927 69I9 8738 9689	0,0463 684I 5008 9268	20
2I	8023 9468 2970 4608	0000 0000 0000 0000	0430 9I88 I434 729I	2I
22	8436 4645 2405 8427	0786 0I58 3579 0968	0393 2789 8624 8I08	22
23	8808 7837 628I I028	0000 0000 0000 0000	0350 5I48 950I 3735	23
24	9I36 0065 7534 8825	0607 5928 5343 95I6	0303 3356 2983 37II	24
25	0,94I4 5733 I626 0285	0,0000 0000 0000 0000	0,0253 4577 I630 2327	25
26	9642 I744 I83I 7868	0400 7904 3579 880I	020I 2529 7436 3443	26
27	98I5 79I6 9I39 4266	0000 0000 0000 0000	0I45 2435 0630 7543	27
28	993I 4I90 4348 4062	0I75 5973 0I65 8759	0085 7422 9454 9678	28
29	9988 6029 6878 27I6	0000 0000 0000 0000	0030 6977 9343 I89I	29
Remainder terms		$G_{28} = 0,167 \times 10^{-6}$	$K_{44} = 0,108 \times 10^{-13}$	

No.	Nodes Mantissa	Order	Weights of Gaussian quadratures Mantissa	Order	Weights of improved quadratures Mantissa	Order	No.
I	4526 1027 2121 5446	-11	0000 0000 0000 0000	0	6222 7131 7050 5376	-10	I
2	7013 4740 2112 4751	- 7	4375 4554 2623 7652	- 5	4307 5354 6143 6467	- 6	2
3	4556 3510 2070 7732	- 5	0000 0000 0000 0000	0	7337 3612 5721 5075	- 6	3
4	4451 0273 2431 4130	- 4	5102 4754 2556 1601	- 4	5115 6716 6237 3301	- 5	4
5	7374 5157 7026 1564	- 4	0000 0000 0000 0000	0	6372 0757 4174 6253	- 5	5
6	5417 1043 3444 2307	- 3	7615 7272 3573 0216	- 4	7607 7027 4612 1672	- 5	6
7	7477 3011 5417 1741	- 3	0000 0000 0000 0000	0	4371 1046 4425 5016	- 4	7
I0	5001 5444 7541 7467	- 2	5017 4735 7561 3227	- 3	5021 3112 3151 7632	- 4	I0
II	6245 4414 2612 0535	- 2	0000 0000 0000 0000	0	5410 0434 6574 1653	- 4	II
I2	7603 0524 5261 0366	- 2	5737 6707 0524 3653	- 3	5736 6316 5007 6673	- 4	I2
I3	4511 0556 2700 6305	- I	0000 0000 0000 0000	0	6223 7437 1305 2415	- 4	I3
I4	5344 7246 6673 2243	- I	6441 7613 6267 6530	- 3	6442 4252 4457 4110	- 4	I4
I5	6220 0203 3076 2176	- I	0000 0000 0000 0000	0	6610 4724 3642 7300	- 4	I5
I6	7105 3202 6222 0611	- I	6706 7040 5250 5161	- 3	6706 5475 5557 3436	- 4	I6
I7	4000 0000 0000 0000	0	0000 0000 0000 0000	0	6733 6262 2543 7463	- 4	I7
20	4335 2276 4666 7473	0	6706 7040 5250 5161	- 3	6706 5475 5557 3436	- 4	20
2I	4667 7676 2340 6700	0	0000 0000 0000 0000	0	6610 4724 3642 7300	- 4	2I
22	5215 4254 4442 2656	0	6441 7613 6267 6530	- 3	6442 4252 4457 4110	- 4	22
23	5533 3510 6437 4635	0	0000 0000 0000 0000	0	6223 7437 1305 2415	- 4	23
24	6087 1652 6523 5702	0	5737 6707 0524 3653	- 3	5736 6316 5007 6673	- 4	24
25	6326 4674 7285 3650	0	0000 0000 0000 0000	0	5410 0434 6574 1653	- 4	25
26	6577 4466 6047 4062	0	5017 4735 7561 3227	- 3	5021 3112 3151 7632	- 4	26
27	7030 0476 6286 0603	0	0000 0000 0000 0000	0	4371 1046 4425 5016	- 4	27
30	7236 0673 4433 3547	0	7615 7272 3573 0216	- 4	7607 7027 4612 1672	- 5	30
3I	7420 1531 0036 4710	0	0000 0000 0000 0000	0	6372 0757 4174 6253	- 5	3I
32	7555 3364 2256 3172	0	5102 4754 2556 1601	- 4	5115 6716 6287 3301	- 5	32
33	7664 4305 5736 1601	0	0000 0000 0000 0000	0	7337 3612 5721 5075	- 6	33
34	7743 7214 1767 3254	0	4375 4554 2623 7652	- 5	4307 5354 6143 6467	- 6	34
35	7773 2516 7505 6562	0	0000 0000 0000 0000	0	6222 7131 7050 5376	-10	35

$$n = 15$$

No.	Nodes	Weights of Gaussian quadratures	Weights of improved quadratures	No.
I	0,0009 9885 0658 30I5	0,0000 0000 0000 0000	0,0026 8873 9936 46I7	I
2	0060 0374 0989 7573	0I53 7662 0998 0586	0075 0397 3664 658I	2
3	0I6I 3046 2I60 4304	0000 0000 0000 0000	0I27 3042 3663 3577	3
4	03I3 6330 3799 6470	035I 8302 3744 054I	0I76 73I8 0395 6879	4
5	05I3 6773 3827 9590	0000 0000 0000 0000	0222 9487 5662 3824	5
6	0,0758 9670 8294 7864	0,0535 796I 0233 5860	0,0267 4076 2345 4640	6
7	I047 9074 9278 7670	0000 0000 0000 0000	03I0 0478 3900 3353	7
8	I377 9II3 43I9 9I50	0697 8533 8963 0772	0349 2706 0659 364I	8
9	I745 0I62 935I 29I5	0000 0000 0000 0000	0384 2484 0378 8602	9
I0	2I45 I39I 3695 7306	083I 3460 2908 4970	04I5 4025 I4II 5665	I0
II	0,2574 5906 8I79 8802	0,0000 0000 0000 0000	0,0442 8222 I528 I059	II
I2	3029 2432 646I 2I83	0930 8050 0007 78II	0465 6329 9085 4I27	I2
I3	3504 0999 6423 4I56	0000 0000 0000 0000	0483 2I36 349I 8II8	I3
I4	3994 0295 300I 2827	0992 I574 2663 5558	0495 8679 9360 8960	I4
I5	4494 2896 6540 64I2	0000 0000 0000 0000	0503 8492 276I 9378	I5
I6	0,5000 0000 0000 0000	0,I0I2 89I2 0962 7806	0,0506 6500 3507 3958	I6
I7	0,5505 7I03 3459 3587	0,0000 0000 0000 0000	0,0503 8492 276I 9378	I7
I8	6005 9704 6998 7I73	0992 I574 2663 5558	0495 8679 9360 8960	I8
I9	6495 9000 3576 5844	0000 0000 0000 0000	0483 2I36 349I 8II8	I9
20	6970 7567 3538 78I7	0930 8050 0007 78II	0465 6329 9085 4I27	20
2I	7425 4093 I820 II98	0000 0000 0000 0000	0442 8222 I528 I059	2I
22	0,7854 8608 6304 2694	0,083I 3460 2908 4970	0,04I5 4025 I4II 5665	22
23	8254 9837 0648 7085	0000 0000 0000 0000	0384 2484 0378 8602	23
24	8622 0886 5680 0850	0697 8533 8963 0772	0349 2706 0659 364I	24
25	8952 0925 072I 2330	0000 0000 0000 0000	03I0 0478 3900 3353	25
26	924I 0329 I705 2I36	0535 796I 0233 5860	0267 4076 2345 4640	26
27	0,9486 3226 6I72 04I0	0,0000 0000 0000 0000	0,0222 9487 5662 3824	27
28	9686 3669 6200 3580	035I 8302 3744 054I	0I76 73I8 0395 6879	28
29	9838 6953 7839 5696	0000 0000 0000 0000	0I27 3042 3663 3577	29
30	9939 9625 90I0 2427	0I53 7662 0998 0586	0075 0397 3664 658I	30
3I	9990 0II4 9346 6985	0000 0000 0000 0000	0026 8873 9936 46I7	3I
Remainder terms		$G_{30} = 0{,}446 \times 10^{-7}$	$K_{48} = 0{,}I27 \times 10^{-I4}$	

32

$$n = 17$$

No.	Nodes Mantissa	Order	Weights of Gaussian quadratures Mantissa	Order	Weights of improved quadratures Mantissa	Order	No.
I	4056 5735 6166 735I	-II	0000 0000 0000 0000	0	5403 2622 0603 22I7	-I0	I
2	6II3 5407 4648 II37	- 7	7676 707I 0475 05II	- 6	7536 I744 5726 4200	- 7	2
3	4I02 2007 7345 455I	- 5	0000 0000 0000 0000	0	64II I504 246I 4030	- 6	3
4	4007 33I6 6035 I356	- 4	440I 6023 0I32 720I	- 4	44I4 3530 364I 5357	- 5	4
5	6446 337I 0334 2577	- 4	0000 0000 0000 0000	0	5552 I676 I655 I602	- 5	5
6	4666 7673 6000 0I23	- 3	6667 3II3 0657 42I2	- 4	6660 756I 3574 6726	- 5	6
7	655I 62I0 22II 585I	- 3	0000 0000 0000 0000	0	7737 6676 4676 I472	- 5	7
I0	432I 4436 3745 2804	- 2	4356 5635 5II3 06I3	- 3	4360 7655 3507 0043	- 4	I0
II	5453 02I6 07I2 3I5I	- 2	0000 0000 0000 0000	0	4726 I535 4307 7745	- 4	II
I2	6672 46II 0367 0306	- 2	5244 II7I 4I24 I566	- 3	5242 3034 2507 6027	- 4	I2
I3	4075 0654 6274 4II6	- I	0000 0000 0000 0000	0	5526 0506 250I 7I73	- 4	I3
I4	466I 4345 5552 0245	- I	5752 0375 I666 7023	- 3	5753 4450 3304 0I35	- 4	I4
I5	5466 4360 3I32 4874	- I	0000 0000 0000 0000	0	6I36 6237 I745 I7I6	- 4	I5
I6	6307 72I3 I642 0766	- I	6263 0637 4376 4026	- 3	626I 5607 062I 4003	- 4	I6
I7	7I4I 56I5 5445 2622	- I	0000 0000 0000 0000	0	6346 0I53 557I 26I2	- 4	I7
20	4000 0000 0000 0000	0	6867 0253 5374 7II4	- 3	6370 3033 2485 5430	- 4	20
2I	43I7 I07I II55 2467	- 0	0000 0000 0000 0000	0	6346 0I53 557I 26I2	- 4	2I
22	4634 0272 3056 7405	0	6263 0637 4376 4026	- 3	626I 5607 062I 4003	- 4	22
23	5I44 5607 6322 5602	0	0000 0000 0000 0000	0	6I36 6237 I745 I7I6	- 4	23
24	5447 I6I5 III2 7655	0	5752 0375 I666 7023	- 3	5753 4450 3304 0I35	- 4	24
25	574I 345I 464I 573I	0	0000 0000 0000 0000	0	5526 0506 250I 7I73	- 4	25
26	622I 2635 5702 I7I6	0	5244 II7I 4I24 I566	- 3	5242 3034 2507 6027	- 4	26
27	6465 I734 36I5 3I46	0	0000 0000 0000 0000	0	4726 I535 4307 7745	- 4	27
30	67I3 4670 3006 53I7	0	4356 5635 5II3 06I3	- 3	4360 7655 3507 0043	- 4	30
3I	7I22 6I56 7556 6243	0	0000 0000 0000 0000	0	7737 6676 4676 I472	- 5	3I
32	73II I0I0 4I77 7765	0	6667 3II3 0657 42I2	- 4	6660 756I 3574 6726	- 5	32
33	7455 4620 3862 I650	0	0000 0000 0000 0000	0	5552 I676 I655 I602	- 5	33
34	7577 4223 0476 I32I	0	440I 6023 0I32 720I	- 4	44I4 3530 364I 5357	- 5	34
35	7675 7337 60I0 6465	0	0000 0000 0000 0000	0	64II I504 246I 4030	- 6	35
36	7747 32II 74I4 5633	0	7676 707I 0475 05II	- 6	7536 I744 5726 4200	- 7	36
37	7773 72I2 042I 6II0	0	0000 0000 0000 0000	0	5403 2622 0603 22I7	-I0	37

No.	Nodes	Weights of Gaussian quadratures	Weights of improved quadratures	No.
I	0,0008 8036 2927 2777	0,0000 0000 0000 0000	0,0023 7I38 8524 6237	I
2	0052 9953 2504 I750	0I35 7622 9705 8770	0066 2896 5344 0456	2
3	0I42 4702 45I5 3037	0000 0000 0000 0000	0II2 4942 9720 0247	3
4	0277 I248 8463 3837	03II 2676 I969 3239	0I56 3027 I823 6903	4
5	0454 2II6 6493 8285	0000 0000 0000 0000	0I97 5647 560I 2II0	5
6	0,067I 8439 8806 084I	0,0475 7925 584I 2464	0,0237 53I0 7988 2035	6
7	0928 7985 6468 7778	0000 0000 0000 0000	0276 028I 6547 7III	7
8	I222 9779 5822 4985	0623 I448 5627 7669	03II 7940 3005 9I74	8
9	I55I 2944 6659 II88	0000 0000 0000 0000	0344 3I49 7595 7656	9
I0	I9I0 6I87 7798 678I	0747 9799 4408 2884	0373 849I I942 7998	I0
II	0,2297 96I6 I823 930I	0,0000 0000 0000 0000	0,0400 2697 063I 8596	II
I2	2709 9I6I II7I 3863	0845 7825 9697 50I3	0422 9790 I896 2953	I2
I3	3I42 58I0 9560 79I9	0000 0000 0000 0000	044I 6875 I289 5564	I3
I4	359I 9822 46I0 3705	09I3 0I70 7522 46I8	0456 460I 64I4 0958	I4
I5	4054 I57I 0490 958I	0000 0000 0000 0000	0467 I933 7030 4606	I5
I6	0,4524 9374 508I I8I3	0,0947 2530 5227 5342	0,0473 6420 0623 6I50	I6
I7	5000 0000 0000 0000	0000 0000 0000 0000	0475 77I0 8040 2492	I7
I8	5475 0625 49I8 8I87	0947 2530 5227 5342	0473 6420 0623 6I50	I8
I9	0,5945 8428 9509 04I9	0,0000 0000 0000 0000	0,0467 I933 7030 4606	I9
20	6408 0I77 5389 6295	09I3 0I70 7522 46I8	0456 460I 64I4 0958	20
2I	6857 4I89 0439 208I	0000 0000 0000 0000	044I 6875 I289 5564	2I
22	7290 0838 8828 6I37	0845 7825 9697 50I3	0422 9790 I896 2953	22
23	7702 0383 8I76 0699	0000 0000 0000 0000	0400 2697 063I 8596	23
24	0,8089 38I2 220I 32I9	0,0747 9799 4408 2884	0,0373 849I I942 7998	24
25	8448 7055 3340 88I2	0000 0000 0000 0000	0344 3I49 7595 7656	25
26	8777 0220 4I77 50I5	0623 I448 5627 7669	03II 7940 3005 9I74	26
27	907I 20I4 353I 2222	0000 0000 0000 0000	0276 028I 6547 7III	27
28	9328 I560 II93 9I59	0475 7925 584I 2464	0237 53I0 7988 2035	28
29	0,9545 7883 3506 I7I5	0,0000 0000 0000 0000	0,0I97 5647 560I 2II0	29
30	9722 875I I586 6I63	03II 2676 I969 3239	0I56 3027 I823 6903	30
3I	9857 5297 5484 6963	0000 0000 0000 0000	0II2 4942 9720 0247	3I
32	9947 0046 7495 8250	0I35 7622 9705 8770	0066 2896 5344 0456	32
33	999I I963 7072 7223	0000 0000 0000 0000	0023 7I38 8524 6237	33
Remainder terms		$G_{32} = 0{,}II9 \times I0^{-7}$	$K_{50} = 0{,}I38 \times I0^{-I5}$	

No.	Nodes		Weights of Gaussian quadratures		Weights of improved quadratures		No.
	Mantissa	Order	Mantissa	Order	Mantissa	Order	
I	7I54 4047 73I5 0665	−I2	0000 0000 0000 0000	0	4666 45I4 I2I4 6I34	−I0	I
2	5332 3663 3I20 7404	− 7	6746 7325 5I33 7020	− 6	6623 37I2 5337 I574	− 7	2
3	7226 6I32 0557 II20	− 6	0000 0000 0000 0000	0	5604 7607 I374 0407	− 6	3
4	7060 25I4 75I0 I450	− 5	7757 66I5 0540 4462	− 5	4000 54I6 2236 7II6	− 5	4
5	5640 56I3 4243 0723	− 4	0000 0000 0000 0000	0	5035 4I25 0456 6360	− 5	5
6	423I 377I 2726 I557	− 3	6056 II67 I7I6 3022	− 4	605I 2740 577I 5450	− 5	6
7	5743 37I3 2376 6262	− 3	0000 0000 0000 0000	0	704I 75I5 2263 II57	− 5	7
I0	7647 3504 2267 0662	− 3	7763 657I 2726 4703	− 4	7766 5762 4I60 I730	− 5	I0
II	4755 5I00 7403 024I	− 2	0000 0000 0000 0000	0	4320 40I2 6040 2222	− 4	II
I2	6072 267I 4452 0564	− 2	4622 7660 6745 0745	− 3	4622 0353 7050 3727	− 4	I2
I3	7264 7657 276I 7323	− 2	0000 0000 0000 0000	0	5077 I522 0722 30I4	− 4	I3
I4	4253 755I 4557 0542	− I	5323 3585 6655 I657	− 3	5324 0220 4525 4025	− 4	I4
I5	50I6 3I60 2742 0775	− I	0000 0000 0000 0000	0	55I6 5II2 7II3 I552	− 4	I5
I6	5576 4324 3I57 3462	− I	5657 6I43 5760 3000	− 3	5657 352I I660 7475	− 4	I6
I7	637I I245 7072 7700	− I	0000 0000 0000 0000	0	5765 6306 4242 7643	− 4	I7
20	7I72 6502 4630 4526	− I	6037 7527 2007 II57	− 3	6040 0366 6255 7656	− 4	20
2I	4000 0000 0000 0000	0	0000 0000 0000 0000	0	6056 0066 5422 I527	− 4	2I
22	4302 4536 5463 5524	0	6037 7527 2007 II57	− 3	6040 0366 6255 7656	− 4	22
23	4603 3255 0342 4037	0	0000 0000 0000 0000	0	5765 6306 4242 7643	− 4	23
24	5I00 5625 63I0 2I46	0	5657 6I43 5760 3000	− 3	5657 352I I660 7475	− 4	24
25	5370 6307 64I6 740I	0	0000 0000 0000 0000	0	55I6 5II2 7II3 I552	− 4	25
26	5652 0II3 I5I0 35I6	0	5323 3585 6655 I657	− 3	5324 0220 4525 4025	− 4	26
27	6I22 6024 I203 4II3	0	0000 0000 0000 0000	0	5077 I522 0722 30I4	− 4	27
30	636I 322I 4665 3642	0	4622 7660 6745 0745	− 3	4622 0353 7050 3727	− 4	30
3I	6604 4557 6077 I727	0	0000 0000 0000 0000	0	4320 40I2 6040 2222	− 4	3I
32	70I3 0427 355I 07II	0	7763 657I 2726 4703	− 4	7766 5762 4I60 I730	− 5	32
33	7203 4406 4540 II5I	0	0000 0000 0000 0000	0	704I 75I5 2263 II57	− 5	33
34	7354 6400 6505 I622	0	6056 II67 I7I6 3022	− 4	605I 2740 577I 5450	− 5	34
35	7505 7507 2I65 6342	0	0000 0000 0000 0000	0	5035 4I25 0456 6360	− 5	35
36	76I6 3725 4605 5746	0	7757 66I5 0540 4462	− 5	4000 54I6 2236 7II6	− 5	36
37	7705 5II6 4572 2066	0	0000 0000 0000 0000	0	5604 7607 I374 0407	− 6	37
40	7752 2260 4623 274I	0	6746 7325 5I33 7020	− 6	6623 37I2 5337 I574	− 7	40
4I	7774 3II5 7540 23I3	0	0000 0000 0000 0000	0	4666 45I4 I2I4 6I34	−I0	4I

$$n = 17$$

No.	Nodes	Weights of Gaussian quadratures	Weights of improved quadratures	No.
I	0,0007 8351 4696 9710	0,0000 0000 0000 0000	0,0021 0948 7896 8885	I
2	0047 1226 2342 7913	0120 7415 1434 2740	0058 9291 8781 1445	2
3	0126 7037 1516 2845	0000 0000 0000 0000	0100 1111 6976 6476	3
4	0246 6223 9115 6161	0277 2976 4686 9936	0139 2836 1228 9317	4
5	0404 5193 1598 0542	0000 0000 0000 0000	0176 2487 3375 9400	5
6	0,0598 8042 3136 5070	0,0425 1807 4158 5896	0,0212 2131 5102 5004	6
7	0828 6295 3574 9328	0000 0000 0000 0000	0247 1807 0911 9795	7
8	1092 4299 8051 5993	0559 4192 3596 7020	0279 9702 2265 0468	8
9	1387 6385 6813 7950	0000 0000 0000 0000	0310 0045 7634 1150	9
I0	1711 6442 0391 6546	0675 6818 4234 2627	0337 6400 8359 0655	I0
II	0,2062 1539 3832 9824	0,0000 0000 0000 0000	0,0362 9494 5057 0951	II
I2	2436 5473 1456 7615	0770 2288 0538 4051	0385 2811 1523 1011	I2
I3	2881 5635 2395 1003	0000 0000 0000 0000	0404 1833 9220 4089	I3
I4	3243 8411 8273 0618	0840 0205 1078 2250	0419 8628 5995 6174	I4
I5	3670 2674 6274 1590	0000 0000 0000 0000	0432 4526 6102 8260	I5
I6	0,4107 5790 9252 0761	0,0882 8135 2683 4963	0,0441 5439 1148 8022	I6
I7	4552 0718 0287 3867	0000 0000 0000 0000	0446 8092 2933 8944	I7
I8	5000 0000 0000 0000	0897 2323 5178 1033	0448 4821 0971 9907	I8
I9	5447 9281 9712 6133	0000 0000 0000 0000	0446 8092 2933 8944	I9
20	5892 4209 0747 9239	0882 8135 2683 4963	0441 5439 1148 8022	20
2I	0,6329 7825 3725 8410	0,0000 0000 0000 0000	0,0432 4526 6102 8260	2I
22	6756 1588 1726 9382	0840 0205 1078 2250	0419 8628 5995 6174	22
23	7168 4364 7604 8997	0000 0000 0000 0000	0404 1833 9220 4089	23
24	7563 4526 8543 2385	0770 2288 0538 4051	0385 2811 1523 1011	24
25	7937 8460 6167 0176	0000 0000 0000 0000	0362 9494 5057 0951	25
26	0,8288 3557 9608 3454	0,0675 6818 4234 2627	0,0337 6400 8359 0655	26
27	8612 3614 3686 2050	0000 0000 0000 0000	0310 0045 7634 1150	27
28	8907 5700 1948 4007	0559 4192 3596 7020	0279 9702 2265 0468	28
29	9171 3704 6425 0672	0000 0000 0000 0000	0247 1807 0911 9795	29
30	9401 1957 6863 4930	0425 1807 4158 5896	0212 2131 5102 5004	30
3I	0,9595 4806 8401 9458	0,0000 0000 0000 0000	0,0176 2487 3375 9400	3I
32	9753 3776 0884 3839	0277 2976 4686 9936	0139 2836 1228 9317	32
33	9873 2962 8483 7155	0000 0000 0000 0000	0100 1111 6976 6476	33
34	9952 8773 7657 2087	0120 7415 1434 2740	0058 9291 8781 1445	34
35	9992 1648 5303 0290	0000 0000 0000 0000	0021 0948 7896 8885	35
Remainder terms		$G_{34} = 0{,}315 \times 10^{-8}$	$K_{54} = 0{,}164 \times 10^{-16}$	

No.	Nodes Mantissa	Order	Weights of Gaussian quadratures Mantissa	Order	Weights of improved quadratures Mantissa	Order	No.
I	6326 2307 7766 4556	-I2	0000 0000 0000 0000	0	4243 7525 4I64 4027	-I0	I
2	4646 4522 2470 I730	- 7	6I35 I25I 3046 03I4	- 6	602I 4542 56I3 207I	- 7	2
3	637I 3543 645I 454I	- 6	0000 0000 0000 0000	0	5I00 2653 00I3 3I4I	- 6	3
4	6240 4I66 6345 I625	- 5	7062 46I0 035I 5357	- 5	7I03 I707 7734 23I2	- 6	4
5	5I33 0354 5506 0365	- 4	0000 0000 0000 0000	0	4406 I0II 6575 6637	- 5	5
6	7524 2454 54I2 57I0	- 4	5342 3556 5000 0057	- 4	5335 4I22 6257 627I	- 5	6
7	5233 20I5 2723 6003	- 3	0000 0000 0000 0000	0	6247 66I5 2I37 647I	- 5	7
I0	6773 53I2 777I 330I	- 3	7I22 I533 6I22 5657	- 4	7I25 5002 7007 0270	- 5	I0
II	434I 4034 6057 007I	- 2	0000 0000 0000 0000	0	7737 2253 7546 I306	- 5	II
I2	5364 267I 6364 4247	- 2	4246 0460 I304 5I03	- 3	4244 6040 7657 0745	- 4	I2
I3	6462 5040 6434 3425	- 2	0000 0000 0000 0000	0	45I2 5002 I007 6I05	- 4	I3
I4	7630 0240 I707 6544	- 2	4733 7054 0I72 307I	- 3	4734 7647 I367 7I46	- 4	I4
I5	44I7 4736 4446 I032	- I	0000 0000 0000 0000	0	5I30 6663 2434 I677	- 4	I5
I6	5I4I 2654 6552 306I	- I	5300 4504 3426 6I25	- 3	5277 47I7 6473 2365	- 4	I6
I7	5676 5356 0000 II65	- I	0000 0000 0000 0000	0	5422 0762 5636 2460	- 4	I7
20	6444 7334 258I 7473	- I	55I4 6332 45I7 5I42	- 3	55I5 5474 0765 3644	- 4	20
2I	722I 0352 3007 5067	- I	0000 0000 0000 0000	0	5560 I527 730I 45I0	- 4	2I
22	4000 0000 0000 0000	0	5574 0320 6I44 00I2	- 3	5573 I30I 7550 5I72	- 4	22
23	4267 36I2 6374 I344	- 0	0000 0000 0000 0000	0	5560 I527 730I 45I0	- 4	23
24	4555 422I 6523 0I42	0	55I4 6332 45I7 5I42	- 3	55I5 5474 0765 3644	- 4	24
25	5040 52I0 7777 7305	0	0000 0000 0000 0000	0	5422 0762 5636 2460	- 4	25
26	53I7 245I 45I2 6347	0	5300 4504 3426 6I25	- 3	5277 47I7 6473 2365	- 4	26
27	5570 I420 5554 7362	0	0000 0000 0000 0000	0	5I30 6663 2434 I677	- 4	27
30	603I 7727 74I6 0246	0	4733 7054 0I72 307I	- 3	4734 7647 I367 7I46	- 4	30
3I	6263 2567 6270 7072	0	0000 0000 0000 0000	0	45I2 5002 I007 6I05	- 4	3I
32	6502 722I 4302 6726	0	4246 0460 I304 5I03	- 3	4244 6040 7657 0745	- 4	32
33	6707 4770 6364 I76I	0	0000 0000 0000 0000	0	7737 2253 7546 I306	- 5	33
34	7I00 4246 5000 6447	0	7I22 I533 6I22 5657	- 4	7I25 5002 7007 0270	- 5	34
35	7254 4576 2505 4I77	0	0000 0000 0000 0000	0	6247 66I5 2I37 647I	- 5	35
36	74I2 5655 I5I7 2503	0	5342 3556 5000 0057	- 4	5335 4I22 6257 627I	- 5	36
37	7532 236I I5I3 4760	0	0000 0000 0000 0000	0	4406 I0II 6575 6637	- 5	37
40	7632 7674 2230 6543	0	7062 46I0 035I 5357	- 5	7I03 I707 7734 23I2	- 6	40
4I	77I4 0642 34I3 2632	0	0000 0000 0000 0000	0	5I00 2653 00I3 3I4I	- 6	4I
42	7754 5455 2665 4370	0	6I35 I25I 3046 03I4	- 6	602I 4542 56I3 207I	- 7	42
43	7774 6246 6340 0045	0	0000 0000 0000 0000	0	4243 7525 4I64 4027	-I0	43

No.	Nodes	Weights of Gaussian quadratures	Weights of improved quadratures	No.
I	0,0007 004I 7293 6586	0,0000 0000 0000 0000	0,00I8 8675 5064 220I	I
2	0042 I74I 5789 5345	0I08 0800 6763 24I7	0052 772I 5283 7723	2
3	0II3 4480 96I2 5695	0000 0000 0000 0000	0089 6668 8I02 8557	3
4	0220 8802 52I4 30II	0248 5727 4447 4849	0I24 8220 I376 8700	4
5	0362 4774 92I3 3283	0000 0000 0000 0000	0I58 I700 9924 4220	5
6	0,0536 9876 675I 222I	0,0382 I286 5I27 4445	0,0I90 7687 9I62 5884	6
7	0743 758I 3283 I4I2	0000 0000 0000 0000	0222 5459 7582 5450	7
8	098I 4752 05I3 7384	0504 7I02 2053 I436	0252 5378 575I 05II	8
9	I248 0975 9I48 32I2	0000 0000 0000 0000	0280 36I9 5I9I 4969	9
I0	I54I 5647 8469 8234	06I2 7760 3355 7392	0306 2678 2895 6793	I0
II	0,I859 98I2 2568 0832	0,0000 0000 0000 0000	0,0330 I997 9096 4085	II
I2	220I I458 4463 0262	0703 2I45 7335 3253	085I 6872 50I2 II22	I2
I3	2562 4547 8707 4608	0000 0000 0000 0000	0370 485I 3690 6575	I3
I4	294I 244I 9268 5787	0773 4233 7563 I326	0386 6603 5082 3I92	I4
I5	3334 8828 8604 296I	0000 0000 0000 0000	0400 I500 9747 548I	I5
I6	0,3740 5688 7I54 2472	0,082I 3824 I872 9I64	0,04I0 7I99 989I 6649	I6
I7	4I55 3I58 5966 7254	0000 0000 0000 0000	04I8 2742 57I8 58I2	I7
I8	4576 I249 3479 I323	0845 7II9 I48I 57I8	0422 8466 6772 0382	I8
I9	0,5000 0000 0000 0000	0,0000 0000 0000 0000	0,0424 3906 9306 3385	I9
20	0,5423 8750 6520 8677	0,0845 7II9 I48I 57I8	0,0422 8466 6772 0382	20
2I	5844 684I 4033 2746	0000 0000 0000 0000	04I8 2742 57I8 58I2	2I
22	6259 43II 2845 7528	082I 3824 I872 9I64	04I0 7I99 989I 6649	22
23	0,6665 II7I I395 7039	0,0000 0000 0000 0000	0,0400 I500 9747 548I	23
24	7058 7558 073I 42I3	0773 4233 7563 I326	0386 6603 5082 3I92	24
25	7437 5452 I292 5392	0000 0000 0000 0000	0370 485I 3690 6575	25
26	7798 854I 5536 9738	0703 2I45 7335 3253	035I 6872 50I2 II22	26
27	8I40 0I87 743I 9I68	0000 0000 0000 0000	0330 I997 9096 4085	27
28	0,8458 4352 I530 I766	0,06I2 7760 3355 7392	0,0306 2678 2895 6793	28
29	875I 9024 085I 6788	0000 0000 0000 0000	0280 36I9 5I9I 4969	29
30	90I8 5247 9486 26I6	0504 7I02 2053 I436	0252 5378 575I 05II	30
3I	9256 2468 67I6 8588	0000 0000 0000 0000	0222 5459 7582 5450	3I
32	9463 0I23 3248 7779	0382 I286 5I27 4445	0I90 7687 9I62 5884	32
33	0,9637 5225 0786 67I7	0,0000 0000 0000 0000	0,0I58 I700 9924 4220	33
34	9779 II97 4785 6989	0248 5727 4447 4849	0I24 8220 I376 8700	34
35	9886 55I9 0387 4305	0000 0000 0000 0000	0089 6668 8I02 8557	35
36	9957 8258 42I0 4655	0I08 0800 6763 24I7	0052 772I 5283 7723	36
37	9992 9958 2706 34I4	0000 0000 0000 0000	00I8 8675 5064 220I	37
Remainder terms		$G_{36} = 0{,}834 \times 10^{-9}$	$K_{56} = 0{,}182 \times 10^{-17}$	

38

No.	Nodes		Weights of Gaussian quadratures		Weights of improved quadratures		No.
	Mantissa	Order	Mantissa	Order	Mantissa	Order	
I	5571 6065 3653 0457	-I2	0000 0000 0000 0000	0	7564 6376 4574 0035	-II	I
2	4243 I077 3436 046I	- 7	542I 2020 7I25 0403	- 6	53I6 6I75 4I66 2I67	- 7	2
3	5635 7624 44I7 5465	- 6	0000 0000 0000 0000	0	4456 4404 0236 730I	- 6	3
4	55I7 0762 I733 357I	- 5	6272 0573 4643 5206	- 5	63I0 I045 53I0 34I5	- 6	4
5	4507 4205 0554 6064	- 4	0000 0000 0000 0000	0	403I I254 426I 0634	- 5	5
6	6677 I474 74I0 I463	- 4	47I0 2427 7002 2056	- 4	4704 3435 407I 7062	- 5	6
7	4605 I025 4427 II37	- 3	0000 0000 0000 0000	0	5544 7506 0632 7574	- 5	7
I0	6220 062I I545 075I	- 3	6353 5263 6435 I27I	- 4	6356 0406 77I5 2240	- 5	I0
II	777I 6I02 234I 6002	- 3	0000 0000 0000 0000	0	7I32 605I 5347 2277	- 5	II
I2	4735 5462 II37 5III	- 2	7657 707I 3I43 7030	- 4	7656 2404 6726 2564	- 5	I2
I3	5747 3II2 5543 0I63	- 2	0000 0000 0000 0000	0	4I63 7765 II25 3277	- 4	I3
I4	7026 2667 555I 0270	- 2	4400 2262 I673 I300	- 3	4400 6424 5704 6774	- 4	I4
I5	4063 I233 378I 3702	- I	0000 0000 0000 0000	0	4574 0056 5252 5443	- 4	I5
I6	455I 357I 6466 6524	- I	4746 2650 6467 0I23	- 3	4746 0I06 6045 0023	- 4	I6
I7	5253 7372 0644 7425	- I	0000 0000 0000 0000	0	5076 3307 3066 7774	- 4	I7
20	5770 2I42 3045 3624	- I	5203 4030 I5I4 I3I2	- 3	5203 5435 0I20 7I47	- 4	20
2I	65I4 02I6 2605 6473	- I	0000 0000 0000 0000	0	5265 I474 06II 4I05	- 4	2I
22	7244 6057 2267 3677	- I	5323 I65I I3I4 3336	- 3	5323 I257 6350 3724	- 4	22
23	4000 0000 0000 0000	0	0000 0000 0000 0000	0	5335 2226 7276 2370	- 4	23
24	4255 4750 2644 2040	0	5323 I65I I3I4 3336	- 3	5323 I257 6350 3724	- 4	24
25	453I 7670 6475 0542	0	0000 0000 0000 0000	0	5265 I474 06II 4I05	- 4	25
26	5003 67I6 6355 2065	0	5203 4030 I5I4 I3I2	- 3	5203 5435 0I20 7I47	- 4	26
27	5252 0202 7455 4I65	0	0000 0000 0000 0000	0	5076 3307 3066 7774	- 4	27
30	55I3 2I03 0544 4525	0	4746 2650 6467 0I23	- 3	4746 0I06 6045 0023	- 4	30
3I	5746 3262 2023 2036	0	0000 0000 0000 0000	0	4574 0056 5252 5443	- 4	3I
32	6I72 3222 0445 572I	0	4400 2262 I673 I300	- 3	4400 6424 5704 6774	- 4	32
33	6406 II55 2447 I743	0	0000 0000 0000 0000	0	4I63 7765 II25 3277	- 4	33
34	66I0 4463 3550 0555	0	7657 707I 3I43 7030	- 4	7656 2404 6726 2564	- 5	34
35	7000 6I67 5543 6I77	0	0000 0000 0000 0000	0	7I32 605I 5347 2277	- 5	35
36	7I55 77I5 6623 2702	0	6353 5263 6435 I27I	- 4	6356 0406 77I5 2240	- 5	36
37	73I7 2675 2335 0664	0	0000 0000 0000 0000	0	5544 7506 0632 7574	- 5	37
40	7444 03I4 I4I7 37I4	0	47I0 2427 7002 2056	- 4	4704 3435 407I 7062	- 5	40
4I	7553 4I67 535I I474	0	0000 0000 0000 0000	0	403I I254 426I 0634	- 5	4I
42	7645 4I60 334I II04	0	6272 0573 4643 5206	- 5	63I0 I045 53I0 34I5	- 6	42
43	772I 420I 5333 6023	0	0000 0000 0000 0000	0	4456 4404 0236 730I	- 6	43
44	7756 5633 402I 6075	0	542I 2020 7I25 0403	- 6	53I6 6I75 4I66 2I67	- 7	44
45	7775 I030 7452 0523	0	0000 0000 0000 0000	0	7564 6376 4574 0035	-II	45

$n = 19$

No.	Nodes	Weights of Gaussian quadratures	Weights of improved quadratures	No.
I	0,0006 3099 3959 8427	0,0000 0000 0000 0000	0,00I6 9907 6559 8I74	I
2	0037 9657 8078 2078	0097 3089 4II4 8632	0047 4960 2730 8550	2
3	0I02 I384 9605 3684	0000 0000 0000 0000	0080 7670 7I84 5457	3
4	0I98 9592 3932 5850	0224 07II 3382 8498	0II2 5434 7876 I02I	4
5	0326 6799 99I7 2403	0000 0000 0000 0000	0I42 72II 9058 5575	5
6	0,0484 2204 8I92 59I0	0,0345 2227 I368 8206	0,0I72 3II7 643I 4405	6
7	067I I322 8657 3324	0000 0000 0000 0000	020I 350I I624 7574	7
8	0886 4267 I73I 4286	0457 450I 08II 2250	0228 9298 6543 96II	8
9	II28 3557 0495 2326	0000 0000 0000 0000	0254 63II 2753 I222	9
I0	I395 I69I I332 3853	0557 8332 2773 6670	0278 76I0 05I6 7407	I0
II	0,I685 3846 7I27 546I	0,0000 0000 0000 0000	0,030I 3870 575I 604I	II
I2	I997 2734 7669 I595	0643 7698 I269 668I	0322 0II7 6I30 9644	I2
I3	2328 62I8 3734 7232	0000 0000 0000 0000	0340 2989 0249 0666	I3
I4	2677 I462 93I2 0I95	07I3 0335 I086 8033	0356 407I 2382 4647	I4
I5	3040 7438 4564 2399	0000 0000 0000 0000	0370 4203 7266 8693	I5
I6	0,34I7 I795 00I8 I85I	0,0763 8302 I032 9298	0,0382 0I43 0969 908I	I6
I7	3803 8753 5840 8793	0000 0000 0000 0000	0390 9345 5342 8709	I7
I8	4I98 2067 7I79 8873	0794 8442 I696 9772	0397 3284 I792 6775	I8
I9	4597 7403 8627 9943	0000 0000 0000 0000	040I 33I0 80I0 4252	I9
20	0,5000 0000 0000 0000	0,0805 2722 4924 39I8	0,0402 7280 I646 499I	20
2I	0,5402 2596 I372 0057	0,0000 0000 0000 0000	0,040I 33I0 80I0 4252	2I
22	580I 7932 2820 II27	0794 8442 I696 9772	0397 3284 I792 6775	22
23	6I96 I246 4I59 I207	0000 0000 0000 0000	0390 9345 5342 8709	23
24	6582 8204 998I 8I49	0763 8302 I032 9298	0382 0I43 0969 908I	24
25	0,6959 256I 5435 760I	0,0000 0000 0000 0000	0,0370 4203 7266 8693	25
26	7322 8537 0687 9805	07I3 0335 I086 8033	0356 407I 2382 4647	26
27	767I 378I 6265 2768	0000 0000 0000 0000	0340 2989 0249 0666	27
28	8002 7265 2330 8405	0643 7698 I269 668I	0322 0II7 6I30 9644	28
29	83I4 6I53 2872 4539	0000 0000 0000 0000	030I 3870 575I 604I	29
30	0,8604 8308 8667 6I47	0,0557 8332 2773 6670	0,0278 76I0 05I6 7407	30
3I	887I 6442 9504 7674	0000 0000 0000 0000	0254 63II 2753 I222	3I
32	9II3 5732 8268 57I4	0457 450I 08II 2250	0228 9298 6543 96II	32
33	9328 8677 I342 6676	0000 0000 0000 0000	020I 350I I624 7574	33
34	95I5 7795 I807 4090	0345 2227 I368 8206	0I72 3II7 643I 4405	34
35	0,9673 3200 0082 7597	0,0000 0000 0000 0000	0,0I42 72II 9058 5575	35
36	980I 0407 6067 4I50	0224 07II 3382 8498	0II2 5434 7876 I02I	36
37	9897 86I5 0394 63I6	0000 0000 0000 0000	0080 7670 7I84 5457	37
38	9962 0342 I92I 7922	0097 3089 4II4 8632	0047 4960 2730 8550	38
39	9993 6900 6040 I573	0000 0000 0000 0000	00I6 9907 6559 8I74	39
Remainder terms		$G_{38} = 0{,}220 \times 10^{-9}$	$K_{60} = 0{,}2I6 \times 10^{-18}$	

No.	Nodes		Weights of Gaussian quadratures		Weights of improved quadratures		No.
	Mantissa	Order	Mantissa	Order	Mantissa	Order	
I	5I26 45II 587I 646I	-I2	0000 0000 0000 0000	0	6753 I6I4 4I00 0I53	-II	I
2	76I5 0002 5353 7672	-I0	4766 7I23 7757 6277	- 6	4672 I2I6 I457 I044	- 7	2
3	5I65 3775 4025 65I4	- 6	0000 0000 0000 0000	0	4I05 2052 2336 I462	- 6	3
4	5057 6306 6472 3452	- 5	5570 7437 3I32 5825	- 5	5606 2050 0I03 I736	- 6	4
5	4I34 734I 3505 I4I7	- 4	0000 0000 0000 0000	0	7235 2633 II47 48I3	- 6	5
6	6I45 3062 4544 3675	- 4	4326 347I 5203 2033	- 4	4822 4I45 3I75 2085	- 5	6
7	4227 I25I 0433 I234	- 3	0000 0000 0000 0000	0	5II7 I056 0335 5I53	- 5	7
I0	5580 5II2 00I6 726I	- 3	5665 7436 6543 5344	- 4	5670 5022 4407 I582	- 5	I0
II	7I6I 3I25 72I5 077I	- 3	0000 0000 0000 0000	0	64II 4004 44I6 6673	- 5	II
I2	4355 6605 3332 2547	- 2	7I07 64I5 537I 50I6	- 4	7I05 6I53 4036 3257	- 5	I2
I3	58II 258I 0433 4436	- 2	0000 0000 0000 0000	0	7556 2562 3430 0I02	- 5	I3
I4	63I0 2523 3226 6I47	- 2	4075 4024 I225 436I	- 3	4076 254I 3I40 0356	- 4	I4
I5	7347 I554 4747 32I6	- 2	0000 0000 0000 0000	0	4266 I355 0664 6243	- 4	I5
I6	422I 0744 246I 6235	- I	4440 3575 620I 5255	- 3	4437 5776 7026 5727	- 4	I6
I7	4672 7643 2050 II54	- I	0000 0000 0000 0000	0	4573 4544 I243 3364	- 4	I7
20	5357 2647 557I 06I3	- I	4706 7263 4247 I353	- 3	4707 4432 4255 06II	- 4	20
2I	6054 I047 5507 6066	- I	0000 0000 0000 0000	0	5002 0I65 4057 3332	- 4	2I
22	6557 I274 2725 405I	- I	5054 4272 3735 3404	- 3	5053 7347 4023 504I	- 4	22
23	7266 3600 5556 5006	- I	0000 0000 0000 0000	0	5I06 I235 II44 20I2	- 4	23
24	4000 0000 0000 0000	0	5II6 5565 I276 3467	- 3	5II7 2427 6775 3423	- 4	24
25	4244 6077 5II0 5375	0	0000 0000 0000 0000	0	5I06 I235 II44 20I2	- 4	25
26	45I0 324I 6425 I753	0	5054 4272 3735 3404	- 3	5053 7347 4023 504I	- 4	26
27	475I 7354 II34 0744	0	0000 0000 0000 0000	0	5002 0I65 4057 3332	- 4	27
30	52I0 2454 II03 3472	0	4706 7263 4247 I353	- 3	4707 4432 4255 06II	- 4	30
3I	5442 4056 2753 73II	0	0000 0000 0000 0000	0	4573 4544 I243 3364	- 4	3I
32	5667 34I5 6547 066I	0	4440 3575 620I 5255	- 3	4437 5776 7026 5727	- 4	32
33	6I06 I444 6606 II34	0	0000 0000 0000 0000	0	4266 I355 0664 6243	- 4	33
34	63I5 7253 II32 2846	0	4075 4024 I225 436I	- 3	4076 254I 3I40 0356	- 4	34
35	65I5 525I 567I 0670	0	0000 0000 0000 0000	0	7556 2562 3430 0I02	- 5	35
36	6704 4236 5III 3246	0	7I07 64I5 537I 50I6	- 4	7I05 6I53 4036 3257	- 5	36
37	706I 6465 2056 2700	0	0000 0000 0000 0000	0	64II 4004 44I6 6673	- 5	37
40	7224 7266 5776 I05I	0	5665 7436 6543 5344	- 4	5670 5022 4407 I582	- 5	40
4I	7355 0652 6734 4654	0	0000 0000 0000 0000	0	5II7 I056 0335 5I53	- 5	4I
42	747I 5234 655I 5604	0	4326 347I 5203 2033	- 4	4822 4I45 3I75 2085	- 5	42
43	7572 I42I 72I3 53I7	0	0000 0000 0000 0000	0	7235 2633 II47 43I3	- 6	43
44	7656 403I 6226 I306	0	5570 7437 3I32 5825	- 5	5606 2050 0I03 I736	- 6	44
45	7726 I240 0237 52I2	0	0000 0000 0000 0000	0	4I05 2052 2336 I462	- 6	45
46	7760 3457 7725 0500	0	4766 7I23 7757 6277	- 6	4672 I2I6 I457 I044	- 7	46
47	7775 3245 533I 2030	0	0000 0000 0000 0000	0	6753 I6I4 4I00 0I53	-II	47

No.	Nodes	Weights of Gaussian quadratures	Weights of improved quadratures	No.
I	0,0005 7048 4205 86I2	0,0000 0000 0000 0000	0,00I5 3679 I859 2608	I
2	0034 3570 0407 4525	0088 0700 3569 576I	0048 00I3 4927 82I5	2
3	0092 4606 I274 8749	0000 0000 0000 0000	0073 I308 4628 4856	3
4	0I80 I403 636I 048I	0203 007I 4900 I935	0I0I 94I8 6730 6333	4
5	0295 8868 3084 I226	0000 0000 0000 0000	0I29 4I06 6802 4756	5
6	0,0438 8278 5874 3370	0,03I3 3602 4I67 0545	0,0I56 4365 3388 5I64	6
7	0608 6I59 4373 8590	0000 0000 0000 0000	0I83 0008 4879 I004	7
8	0804 4I5I 4088 8906	04I6 3837 0788 3524	0208 3443 6663 9868	8
9	I024 7928 558I 2224	0000 0000 0000 0000	0232 I74I 0933 7488	9
I0	I268 3404 6769 9246	0509 6505 9908 6202	0254 7228 696I 8643	I0
II	0,I533 8II7 I832 6243	0,0000 0000 0000 0000	0,0275 9755 2674 I430	II
I2	I8I9 73I5 9636 7425	0590 9726 5980 7592	0295 5570 0440 3I98	I2
I3	2I24 2977 6590 I448	0000 0000 0000 0000	03I3 266I 8777 3906	I3
I4	2445 6649 9024 5864	0658 443I 9224 5883	0329 I729 8566 8092	I4
I5	2782 034I 2380 6374	0000 0000 0000 0000	0343 2433 6464 2608	I5
I6	0,3I3I 4695 5642 2902	0,07I0 4805 4659 I9I0	0,0355 272I I776 7220	I6
I7	349I 8606 5942 5435	0000 0000 0000 0000	0365 I584 5I66 3933	I7
I8	386I 0707 4429 I775	0745 8649 3236 30I9	0372 9I43 7700 2496	I8
I9	4236 9726 7379 5387	0000 0000 0000 0000	0378 5224 8842 2783	I9
20	46I7 3673 9433 25I3	0763 7669 3565 3629	038I 8893 3836 0404	20
2I	0,5000 0000 0000 0000	0,0000 0000 0000 0000	0,0383 0085 5958 9998	2I
22	0,5382 6326 0566 7487	0,0763 7669 3565 3629	0,038I 8893 3836 0404	22
23	5763 0273 2620 46I3	0000 0000 0000 0000	0378 5224 8842 2783	23
24	6I38 9292 5570 8225	0745 8649 3236 30I9	0372 9I43 7700 2496	24
25	6508 I393 4057 4565	0000 0000 0000 0000	0365 I534 5I66 3933	25
26	6868 5304 4357 7098	07I0 4805 4659 I9I0	0355 272I I776 7220	26
27	0,72I7 9658 76I9 3626	0,0000 0000 0000 0000	0,0343 2433 6464 2608	27
28	7554 3850 0975 4I35	0658 443I 9224 5883	0329 I729 8566 8092	28
29	7875 7022 3409 8552	0000 0000 0000 0000	03I3 266I 8777 3906	29
30	8I80 2684 0363 2575	0590 9726 5980 7592	0295 5570 0440 3I98	30
3I	8466 I882 8I67 3757	0000 0000 0000 0000	0275 9755 2674 I430	3I
32	0,873I 6595 3230 0754	0,0509 6505 9908 6202	0,0254 7228 696I 8643	32
33	8975 207I 44I8 7756	0000 0000 0000 0000	0232 I74I 0933 7488	33
34	9I95 5848 59II I094	04I6 3837 0788 3524	0208 3443 6663 9868	34
35	939I 3840 5626 I4I0	0000 0000 0000 0000	0I83 0008 4879 I004	35
36	956I I72I 4I25 6630	03I3 3602 4I67 0545	0I56 4365 3388 5I64	36
37	0,9704 II3I 69I5 8774	0,0000 0000 0000 0000	0,0I29 4I06 6802 4756	37
38	98I9 8596 3638 9569	0203 007I 4900 I935	0I0I 94I8 6730 6333	38
39	9907 5393 8725 I25I	0000 0000 0000 0000	0073 I308 4628 4856	39
40	9965 6429 9592 5475	0088 0700 3569 576I	0043 00I3 4927 82I5	40
4I	9994 295I 5794 I388	0000 0000 0000 0000	00I5 3679 I859 2603	4I
Remainder terms		$G_{40} = 0{,}579 \times 10^{-10}$	$K_{62} = 0{,}243 \times 10^{-19}$	

42

$$n = 24$$

No.	Nodes Mantissa	Order	Weights of Gaussian quadratures Mantissa	Order	Weights of improved quadratures Mantissa	Order	No.
I	4530 6214 0168 5107	-12	0000 0000 0000 0000	0	6226 7055 4827 2321	-11	I
2	7022 4574 7070 6176	-10	4404 5500 0472 3505	-6	4816 4045 3421 1285	-7	2
3	4567 6312 5453 3254	-6	0000 0000 0000 0000	0	7872 1231 5177 3406	-7	3
4	4471 1054 1024 2260	-5	5144 6657 2451 6663	-5	5160 2604 5170 1504	-6	4
5	7446 1767 2133 1114	-5	0000 0000 0000 0000	0	6500 8304 5804 4566	-6	5
6	5473 7157 5060 0233	-4	4005 5063 7373 7443	-4	4002 3436 3543 1107	-5	6
7	7624 5001 6558 0265	-4	0000 0000 0000 0000	0	4536 5017 2043 5572	-5	7
10	5113 7205 2040 2443	-3	5250 6377 0313 3422	-4	5252 6373 0064 3352	-5	10
11	6436 0250 7013 2736	-3	0000 0000 0000 0000	0	5743 1160 4521 0125	-5	11
12	4036 0310 6271 2653	-2	6414 0275 0613 6501	-4	6412 5501 7351 1675	-5	12
13	4720 7764 1465 7604	-2	0000 0000 0000 0000	0	7041 2103 2136 6060	-5	13
14	5645 3454 0130 7362	-2	7440 7771 4253 4730	-4	7441 7313 6632 3104	-5	14
15	6630 3460 7711 5343	-2	0000 0000 0000 0000	0	4005 0127 1444 5131	-4	15
16	7646 7643 7300 4000	-2	4155 4542 7331 1557	-3	4155 2112 0321 1345	-4	16
17	4347 0255 3671 7361	-1	0000 0000 0000 0000	0	4311 3654 7222 1150	-4	17
20	5005 2314 1652 3264	-1	4430 0644 3632 2451	-3	4430 2373 2262 7566	-4	20
21	5454 4204 0656 1136	-1	0000 0000 0000 0000	0	4531 0435 2512 5502	-4	21
22	6132 7723 4455 5350	-1	4614 0315 5226 1314	-3	4613 7350 0527 7212	-4	22
23	6616 7331 1131 4560	-1	0000 0000 0000 0000	0	4660 5365 5333 4344	-4	23
24	7306 4302 0147 2171	-1	4706 5542 2205 0633	-3	4706 5777 6745 0603	-4	24
25	4000 0000 0000 0000	0	0000 0000 0000 0000	0	4716 0325 4104 7024	-4	25
26	4234 5636 7714 2703	0	4706 5542 2205 0633	-3	4706 5777 6745 0603	-4	26
27	4470 4228 3323 1507	0	0000 0000 0000 0000	0	4660 5365 5333 4344	-4	27
30	4722 4026 1551 1213	0	4614 0315 5226 1314	-3	4613 7350 0527 7212	-4	30
31	5151 5675 7450 7320	0	0000 0000 0000 0000	0	4531 0435 2512 5502	-4	31
32	5375 2631 7052 6245	0	4430 0644 3632 2451	-3	4430 2373 2262 7566	-4	32
33	5614 3651 2043 0207	0	0000 0000 0000 0000	0	4311 3654 7222 1150	-4	33
34	6026 2027 0117 6777	0	4155 4542 7331 1557	-3	4155 2112 0321 1345	-4	34
35	6231 7063 6015 4507	0	0000 0000 0000 0000	0	4005 0127 1444 5131	-4	35
36	6426 5064 7751 6103	0	7440 7771 4253 4732	-4	7441 7313 6632 3104	-5	36
37	6613 6002 7462 4036	0	0000 0000 0000 0000	0	7041 2103 2136 6060	-5	37
40	6770 3715 6321 5225	0	6414 0275 0613 6501	-4	6412 5501 7351 1675	-5	40
41	7134 1752 7076 4504	0	0000 0000 0000 0000	0	5743 1160 4521 0125	-5	41
42	7266 4057 2573 7533	0	5250 6377 0313 3422	-4	5252 6373 0064 3352	-5	42
43	7406 5537 7051 2364	0	0000 0000 0000 0000	0	4536 5017 2043 5572	-5	43
44	7514 2031 0134 7766	0	4005 5063 7373 7443	-4	4002 3436 3543 1107	-5	44
45	7606 6340 2135 1155	0	0000 0000 0000 0000	0	6500 8304 5804 4566	-6	45
46	7666 1556 4757 2732	0	5144 6657 2451 6663	-5	5160 2604 5170 1504	-6	46
47	7732 1014 6523 2445	0	0000 0000 0000 0000	0	7872 1231 5177 3406	-7	47
50	7761 7326 4061 6163	0	4404 5500 0472 3505	-6	4816 4045 3421 1235	-7	50
51	7775 5234 6717 7061	0	0000 0000 0000 0000	0	6226 7055 4827 2321	-11	51

$$n = 21$$

No.	Nodes	Weights of Gaussian quadratures	Weights of improved quadratures	No.
I	0,0005 I903 4656 84I4	0,0000 0000 0000 0000	0,00I3 9774 06I6 2058	I
2	003I 239I 4689 8052	0080 086I 4I28 8872	0039 09I6 865I 062I	2
3	0084 0756 82I9 I570	0000 0000 0000 0000	0066 5232 I303 5442	3
4	0I68 8658 07I6 8469	0I84 7689 4885 4262	0092 8002 7704 438I	4
5	0269 2689 I6I0 7487	0000 0000 0000 0000	0II7 8709 954I 5I45	5
6	0,0399 5033 2924 7996	0,0285 672I 27I3 4286	0,0I42 592I 3285 I025	6
7	0554 3584 2334 4078	0000 0000 0000 0000	0I67 0I27 3I60 4038	7
8	0733 I83I 7708 34I4	0380 5005 68I4 I896	0I90 4I57 6726 7325	8
9	0934 7370 9576 2840	0000 0000 0000 0000	02I2 4846 6440 5592	9
I0	II57 800I 8262 I6I0	0467 222I I728 0I69	0233 4866 2365 7572	I0
II	0,I40I 3827 37I3 9876	0,0000 0000 0000 0000	0,0253 4859 8898 2588	II
I2	I664 3059 790I 2938	0543 9864 9583 5742	0272 0933 7589 I9I8	I2
I3	I945 0I43 8I05 II70	0000 0000 0000 0000	0289 0522 8748 8048	I3
I4	224I 9058 2056 390I	0609 I570 8026 8643	0304 4934 0I02 33I0	I4
I5	2553 5204 5552 I657	0000 0000 0000 0000	0318 4780 7624 5504	I5
I6	0,2878 2893 9896 2806	0,066I 3446 93I6 6687	0,0330 7478 8907 9756	I6
I7	32I4 3787 3I70 0895	0000 0000 0000 0000	034I I035 3I30 5I26	I7
I8	3559 8934 I598 7994	0699 4369 7395 5366	0349 6488 2504 I994	I8
I9	39I3 0995 3064 I356	0000 0000 0000 0000	0356 4667 3I8I 8923	I9
20	4272 I907 29I9 5524	0722 6220 I994 9850	036I 3774 IIII 0I70	20
2I	0,4635 I637 6202 I047	0,0000 0000 0000 0000	0,0364 2282 64I5 2542	2I
22	5000 0000 0000 0000	0730 4056 6824 8452	0365 I374 398I 3845	22
23	5364 8362 3797 8953	0000 0000 0000 0000	0364 2282 64I5 2542	23
24	0,5727 8092 7080 4475	0,0722 6220 I994 9850	0,036I 3774 IIII 0I70	24
25	6086 9004 6935 8644	0000 0000 0000 0000	0356 4667 3I8I 8923	25
26	6440 I065 840I 2005	0699 4369 7395 5366	0349 6488 2504 I994	26
27	6785 6262 6829 9605	0000 0000 0000 0000	034I I035 3I30 5I26	27
28	7I2I 7I06 0I03 7I94	066I 3446 93I6 6687	0330 7478 8907 9756	28
29	0,7446 4795 4447 8343	0,0000 0000 0000 0000	0,03I8 4780 7624 5504	29
30	7758 094I 7943 6099	0609 I570 8026 8643	0304 4934 0I02 33I0	30
3I	8054 9856 I894 8830	0000 0000 0000 0000	0289 0522 8748 8048	3I
32	8335 6940 2098 7062	0543 9864 9583 5742	0272 0933 7589 I9I8	32
33	8598 6I72 6286 0I24	0000 0000 0000 0000	0253 4859 8898 2588	33
34	0,8842 I998 I737 8390	0,0467 222I I728 0I69	0,0233 4866 2365 7572	34
35	9065 2629 0423 7660	0000 0000 0000 0000	02I2 4846 6440 5592	35
36	9266 8I68 229I 6586	0380 5005 68I4 I896	0I90 4I57 6726 7325	36
37	9445 64I5 7665 5922	0000 0000 0000 0000	0I67 0I27 3I60 4038	37
38	9600 4966 7075 2004	0285 672I 27I3 4286	0I42 592I 3285 I025	38
39	0,9730 7360 8389 2563	0,0000 0000 0000 0000	0,0II7 8709 954I 5I45	39
40	9836 I34I 9283 I53I	0I84 7689 4885 4262	0092 8002 7704 438I	40
4I	99I5 9243 I780 8430	0000 0000 0000 0000	0066 5232 I303 5442	4I
42	9968 7608 53I0 I948	0080 086I 4I28 8872	0039 09I6 865I 062I	42
43	9994 8096 5343 6586	0000 0000 0000 0000	00I3 9774 06I6 2058	43
Remainder terms		$G_{42} = 0{,}152 \times 10^{-10}$	$K_{66} = 0{,}290 \times 10^{-20}$	

44

No.	Nodes		Weights of Gaussian quadratures		Weights of improved quadratures		No.
	Mantissa	Order	Mantissa	Order	Mantissa	Order	
I	4200 7723 3767 3342	-I2	0000 0000 0000 0000	0	5563 2I44 3604 2037	-II	I
2	63I3 5227 333I 7402	-I0	4063 32I7 7424 25II	- 6	400I 4I73 6026 I735	- 7	2
3	4233 7745 505I I036	- 6	0000 0000 0000 0000	0	6637 5667 I607 4562	- 7	3
4	4I43 6446 42I3 2470	- 5	4565 6333 3207 3422	- 5	4600 550I 6774 3475	- 6	4
5	67II 2274 3274 I002	- 5	0000 0000 0000 0000	0	602I 7255 6056 4576	- 6	5
6	5072 I365 6540 2556	- 4	7240 27II 4257 I025	- 5	723I 757I 5347 5037	- 6	6
7	706I 026I 4732 I720	- 4	0000 0000 0000 0000	0	42I5 0433 60I5 0336	- 5	7
I0	4542 3752 0I40 5I36	- 3	4675 5I40 2640 5722	- 4	4677 6424 5234 3546	- 5	I0
II	5766 7444 7325 I550	- 3	0000 0000 0000 0000	0	534I 0503 4347 4662	- 5	II
I2	732I 7022 7773 5850	- 3	5765 77I2 I536 2736	- 4	5764 266I 52I0 25I7	- 5	I2
I3	4370 0I50 264I I760	- 2	0000 0000 0000 0000	0	6872 3735 32I0 7344	- 5	I3
I4	5246 63I0 2702 4274	- 2	6755 0436 2350 3472	- 4	6756 8085 7056 6304	- 5	I4
I5	6I62 5542 4353 7853	- 2	0000 0000 0000 0000	0	73I4 5250 4I2I 5300	- 5	I5
I6	7I3I I067 2200 I573	- 2	7630 I277 67II 5753	- 4	7627 0344 773I I7I6	- 5	I6
I7	4053 6600 6623 4774	- I	0000 0000 0000 0000	0	4047 I330 6062 2734	- 4	I7
20	4465 7I20 4566 23I4	- I	4I67 0602 I555 I707	- 3	4I67 4556 0256 4344	- 4	20
2I	5III I560 3657 2I66	- I	0000 0000 0000 0000	0	4273 35I4 I446 4264	- 4	2I
22	5544 2074 II74 4I47	- I	4363 7244 II60 0II6	- 3	4363 3526 I335 I405	- 4	22
23	6205 4707 I540 2376	- I	0000 0000 0000 0000	0	4440 I076 7607 0I27	- 4	23
24	6653 6I65 25I5 2566	- I	4477 7064 4447 057I	- 3	4500 2453 007I 0763	- 4	24
25	7325 I004 5636 2636	- I	0000 0000 0000 0000	0	4523 0032 0II0 0237	- 4	25
26	4000 0000 0000 0000	0	453I 3II2 7345 5043	- 3	4530 7557 4I0I I26I	- 4	26
27	4225 3375 5060 6460	0	0000 0000 0000 0000	0	4523 0032 0II0 0237	- 4	27
30	4452 0705 253I 2504	0	4477 7064 4447 057I	- 3	4500 2453 007I 0763	- 4	30
3I	4675 I434 3II7 6600	0	0000 0000 0000 0000	0	4440 I076 7607 0I27	- 4	3I
32	5II5 674I 730I 57I4	0	4363 7244 II60 0II6	- 3	4363 3526 I335 I405	- 4	32
33	5333 3I07 6050 2704	0	0000 0000 0000 0000	0	4273 35I4 I446 4264	- 4	33
34	5545 0327 5504 663I	0	4I67 0602 I555 I707	- 3	4I67 4556 0256 4344	- 4	34
35	5752 0477 4466 I40I	0	0000 0000 0000 0000	0	4047 I330 6062 2734	- 4	35
36	6I5I 5562 I337 744I	0	7630 I277 67II 5753	- 4	7627 0344 773I I7I6	- 5	36
37	6343 2447 2705 0I05	0	0000 0000 0000 0000	0	73I4 5250 4I2I 5300	- 5	37
40	6526 23I5 72I7 2720	0	6755 0436 2350 3472	- 4	6756 3035 7056 6304	- 5	40
4I	670I 7745 7227 5403	0	0000 0000 0000 0000	0	6372 3735 32I0 7344	- 5	4I
42	7045 6075 5000 4242	0	5765 77I2 I536 2736	- 4	5764 266I 52I0 25I7	- 5	42
43	720I I033 3045 2622	0	0000 0000 0000 0000	0	534I 0503 4347 4662	- 5	43
44	7323 5402 5763 7264	0	4675 5I40 2640 5722	- 4	4677 6424 5234 3546	- 5	44
45	7434 7364 7I42 2702	0	0000 0000 0000 0000	0	42I5 0433 60I5 0336	- 5	45
46	7534 2720 505I 765I	0	7240 27II 4257 I025	- 5	723I 757I 5347 5037	- 6	46
47	762I 5582 07I2 0757	0	0000 0000 0000 0000	0	602I 7255 6056 4576	- 6	47
50	7674 7026 6273 5I26	0	4565 6333 3207 3422	- 5	4600 550I 6774 3475	- 6	50
5I	7735 4400 3227 2667	0	0000 0000 0000 0000	0	6637 5667 I607 4562	- 7	5I
52	7763 I505 32II II40	0	4063 32I7 7424 25II	- 6	400I 4I73 6026 I735	- 7	52
53	7775 6774 0262 0042	0	0000 0000 0000 0000	0	5563 2I44 3604 2037	-II	53

No.	Nodes	Weights of Gaussian quadratures	Weights of improved quadratures	No.
I	0,0004 736I 3803 9098	0,0000 0000 0000 0000	0,00I2 7586 6646 I056	I
2	0028 5270 7258 8004	0073 I399 7649 I36I	0035 7II2 2886 5644	2
3	0076 7963 920I 3258	0000 0000 0000 0000	0060 7735 7267 808I	3
4	0I49 6975 I082 2856	0I68 8745 0792 407I	0084 8023 6406 I799	4
5	0246 0480 9828 4075	0000 0000 0000 0000	0I07 7974 7224 3053	5
6	0,0365 2I6I 3906 4I30	0,026I 4666 7576 34I6	0,0I30 5293 5588 I5I5	6
7	0507 0307 39I0 I940	0000 0000 0000 0000	0I53 0080 06II 4548	7
8	0670 937I II39 8499	0348 9823 42I2 2602	0I74 6200 0230 39I2	8
9	0855 8989 7764 2846	0000 0000 0000 0000	0I95 I350 3740 4I75	9
I0	I060 9I59 70I0 3959	0429 7080 3I08 5339	02I4 7662 7432 5588	I0
II	0,I285 I440 3060 496I	0,0000 0000 0000 0000	0,0233 52I6 2479 6940	II
I2	I527 5636 8406 6586	0502 0707 222I 4405	025I 0972 8I47 II83	I2
I3	I786 887I 3703 5438	0000 0000 0000 0000	0267 3286 7420 6033	I3
I4	206I 7979 8246 5442	0564 66I4 8040 2696	0282 2864 8930 4403	I4
I5	235I 0330 0684 7425	0000 0000 0000 0000	0295 9559 I5I4 56I3	I5
I6	0,2653 2208 I006 62I5	0,06I6 26I8 8405 2562	0,0308 I620 6569 429I	I6
I7	2966 8344 7282 8942	0000 0000 0000 0000	03I8 8056 5I45 0822	I7
I8	3290 3208 9553 9579	0655 8675 2393 58I2	0327 9I30 5I94 8747	I8
I9	3622 I427 4327 I924	0000 0000 0000 0000	0335 4637 2I66 2083	I9
20	3960 6978 6655 8894	0682 7074 9I73 0076	034I 3656 I754 7982	20
2I	0,4304 3II0 4638 5I9I	0,0000 0000 0000 0000	0,0345 5809 9709 8328	2I
22	465I 3036 3340 I389	0696 2593 6427 8I60	0348 I258 II82 5659	22
23	5000 0000 0000 0000	0000 0000 0000 0000	0348 9822 350I 7089	23

No.	Nodes Mantissa	Order	Weights of Gaussian quadratures Mantissa	Order	Weights of improved quadratures Mantissa	Order	No.
I	7604 7536 3074 5352	−I3	0000 0000 0000 0000	0	5I63 5873 0234 2363	−II	I
2	5657 2I74 3033 0446	−I0	7372 5I02 2646 0060	− 7	7240 4600 0037 7236	−I0	2
3	7672 2573 4667 6664	− 7	0000 0000 0000 0000	0	6I62 222I 2660 65I6	− 7	3
4	7524 I657 65I7 4252	− 6	4245 36I5 0673 3744	− 5	4257 0260 4033 4226	− 6	4
5	623I 0006 5423 0643	− 5	0000 0000 0000 0000	0	54II 66II 3446 3406	− 6	5
6	453I 3660 0523 I762	− 4	6543 06II 2I07 4400	− 5	6535 5772 6737 I05I	− 6	6
7	6372 7006 6I00 I750	− 4	0000 0000 0000 0000	0	7653 0065 4352 2350	− 6	7
I0	4226 4I55 36I6 3505	− 3	4857 0563 324I I657	− 4	4360 6I67 7723 0532	− 5	I0
II	5364 470I 474I 7753	− 3	0000 0000 0000 0000	0	4775 53I0 48I2 3652	− 5	II
I2	6624 32I5 0705 I0I3	− 3	5400 I047 I0I0 55I4	− 4	5876 7700 4347 0I35	− 5	I2
I3	407I 4507 46I5 3324	− 2	0000 0000 0000 0000	0	5764 64I0 6II0 2I52	− 5	I3
I4	4706 6052 2643 7722	− 2	6332 2756 2470 0402	− 4	6333 I352 5I50 0640	− 5	I4
I5	5557 5054 454I 26I0	− 2	0000 0000 0000 0000	0	6657 7342 723I 3753	− 5	I5
I6	6462 08I4 0244 II27	− 2	7I64 44I4 I45I 7765	− 4	7I63 7704 3756 6636	− 5	I6
I7	74I3 7353 3347 662I	− 2	0000 0000 0000 0000	0	7447 II64 I522 2547	− 5	I7
20	4I75 4II3 5632 4403	− I	7706 5675 7707 0577	− 4	7707 II04 73I6 0432	− 5	20
2I	4576 3344 4353 0I53	− I	0000 0000 0000 0000	0	405I 2462 0255 0550	− 4	2I
22	5207 3344 6777 0005	− I	4I45 II30 7II2 3006	− 3	4I45 0054 7547 5565	− 4	22
23	5627 2046 I7I4 II43	− I	0000 0000 0000 0000	0	4226 3753 5532 3447	− 4	23
24	6254 4650 5623 672I	− I	4275 06I0 6574 2632	− 3	4275 I307 4I25 5433	− 4	24
25	6706 0567 I663 24I2	− I	0000 0000 0000 0000	0	4330 63I3 2020 6333	− 4	25
26	7342 262I II35 4I2I	− I	435I 40I2 7764 4550	− 3	435I 3643 0653 I0II	− 4	26
27	4000 0000 0000 0000	0	0000 0000 0000 0000	0	4357 0560 4240 6654	− 4	27

No.	Nodes	Weights of Gaussian quadratures	Weights of improved quadratures	No.
24	5348 6963 6659 86II	0696 2593 6427 8I60	0348 I258 II82 5659	24
25	5695 6889 536I 4809	0000 0000 0000 0000	0345 5809 9709 8328	25
26	0,6039 302I 3344 II06	0,0682 7074 9I73 0076	0,034I 3656 I754 7982	26
27	6377 8572 5672 8076	0000 0000 0000 0000	0335 4637 2I66 2083	27
28	6709 679I 0446 042I	0655 8675 2393 53I2	0327 9I30 5I94 8747	28
29	7033 I655 27I7 I058	0000 0000 0000 0000	03I8 8056 5I45 0822	29
30	7346 779I 8998 3785	06I6 26I8 8405 2562	0308 I620 6569 429I	30
3I	0,7648 9669 93I5 2575	0,0000 0000 0000 0000	0,0295 9559 I5I4 56I3	3I
32	7988 2020 I753 4558	0564 66I4 8040 2696	0282 2864 8930 4403	32
33	82I3 II28 6296 4562	0000 0000 0000 0000	0267 3286 7420 6033	33
34	8472 4363 I593 34I4	0502 0707 222I 4405	025I 0972 8I47 II83	34
35	87I4 8559 6939 5039	0000 0000 0000 0000	0233 52I6 2479 6940	35
36	0,8939 0840 2989 604I	0,0429 7080 3I08 5339	0,02I4 7662 7432 5588	36
37	9I44 I0I0 2285 7I54	0000 0000 0000 0000	0I95 I350 3740 4I75	37
38	9329 0628 8860 I50I	0348 9823 42I2 2602	0I74 6200 0230 39I2	38
39	9492 9692 6089 8060	0000 0000 0000 0000	0I58 0080 06II 4548	39
40	9634 7838 6093 5870	026I 4666 7576 34I6	0I30 5293 5588 I5I5	40
4I	0,9753 95I9 0I7I 5925	0,0000 0000 0000 0000	0,0I07 7974 7224 3053	4I
42	9850 3024 89I7 7I44	0I68 8745 0792 407I	0084 8023 6406 I799	42
43	9923 2036 0798 6742	0000 0000 0000 0000	0060 7735 7267 808I	43
44	997I 4729 274I I996	0073 I399 7649 I36I	0035 7II2 2886 5644	44
45	9995 2638 6I96 0902	0000 0000 0000 0000	00I2 7586 6646 I056	45
Remainder terms		$G_{44} = 0,397 \times 10^{-II}$	$K_{68} = 0,330 \times 10^{-2I}$	

No.	Nodes		Weights of Gaussian quadratures		Weights of improved quadratures		No.
	Mantissa	Order	Mantissa	Order	Mantissa	Order	
30	42I6 6467 332I I727	0	435I 40I2 7764 4550	− 3	435I 3643 0653 I0II	− 4	30
3I	4434 7504 3046 2572	0	0000 0000 0000 0000	.0	4330 63I3 2020 6333	− 4	3I
32	465I 5453 5066 0427	0	4275 06I0 6574 2632	− 3	4275 I307 4I25 5433	− 4	32
33	5064 2754 703I 73I6	0	0000 0000 0000 0000	0	4226 3753 5532 3447	− 4	33
34	5274 22I5 4400 3775	0	4I45 II30 7II2 3006	− 3	4I45 0054 7547 5565	− 4	34
35	5500 62I5 56I2 37I2	0	0000 0000 0000 0000	0	405I 2462 0255 0550	− 4	35
36	570I I732 I062 5576	0	7706 5675 7707 0577	− 4	7707 II04 73I6 0432	− 5	36
37	6075 0I05 II06 0233	0	0000 0000 0000 0000	0	7447 II64 I522 2547	− 5	37
40	6263 37I4 7726 7552	0	7I64 44I4 I45I 7765	− 4	7I63 7704 3756 6636	− 5	40
4I	6444 0564 6647 5235	0	0000 0000 0000 0000	0	6657 7342 723I 3753	− 5	4I
42	66I6 2365 3227 00I3	0	6332 2756 2470 0402	− 4	6333 I352 5I50 0640	− 5	42
43	676I 4656 0634 5II2	0	0000 0000 0000 0000	0	5764 64I0 6II0 2I52	− 5	43
44	7II5 3456 2707 2676	0	5400 I047 I0I0 55I4	− 4	5376 7700 4347 0I35	− 5	44
45	724I 3307 6303 6002	0	0000 0000 0000 0000	0	4775 58I0 48I2 3652	− 5	45
46	7355 I862 24I6 I427	0	4357 0563 324I I657	− 4	4360 6I67 7723 0532	− 5	46
47	7460 2437 4473 770I	0	0000 0000 0000 0000	0	7653 0065 4852 2350	− 6	47
50	7552 3204 7752 6300	0	6543 06II 2I07 4400	− 5	6535 5772 6737 I05I	− 6	50
5I	7633 I577 6247 3I62	0	0000 0000 0000 0000	0	54II 66II 3446 3406	− 6	5I
52	7702 536I 20I2 6035	0	4245 36I5 0673 3744	− 5	4257 0260 4033 4226	− 6	52
53	7740 4265 02I4 4404	0	0000 0000 0000 0000	0	6I62 222I 2660 65I6	− 7	53
54	7764 24I3 407I 7II6	0	7372 5I02 2646 0060	− 7	7240 4600 0037 7236	−I0	54
55	7776 0366 0503 I606	0	0000 0000 0000 0000	0	5I63 5373 0234 2363	−II	55

No.	Nodes	Weights of Gaussian quadratures	Weights of improved quadratures	No.
I	0,0004 3443 I768 2893	0,0000 0000 0000 0000	0,00II 6999 4749 7333	I
2	0026 I533 250I 2239	0067 0592 9743 5709	0032 7342 9700 698I	2
3	0070 4092 4636 3I22	0000 0000 0000 0000	0055 7349 094I 4I76	3
4	0I87 2876 4390 9424	0I54 9400 2928 4897	0077 8I67 8730 5560	4
5	0225 7208 0069 9444	0000 0000 0000 0000	0098 9579 99I4 0407	5
6	0,0335 I445 6586 99I9	0,0240 I883 5865 5423	0,0II9 89I5 7930 69I4	6
7	0465 46I5 2629 3593	0000 0000 0000 0000	0I40 67I9 5372 29I3	7
8	06I6 2382 0864 7792	032I I62I 0704 2629	0I60 7I77 05I2 7460	8
9	0786 5633 850I 4935	0000 0000 0000 0000	0I79 7828 9728 8304	9
I0	0975 5579 9I90 580I	0396 4070 5888 3595	0I98 I0I5 6800 3308	I0
II	0,II82 5434 I889 4I73	0,0000 0000 0000 0000	0,02I5 7347 I9I2 0782	II
I2	I406 693I 8434 0249	0464 5788 3030 0I76	0232 3706 6585 6244	I2
I3	I646 88I0 548I 4788	0000 0000 0000 0000	0247 809I I975 0I02	I3
I4	I90I 9506 2II8 I769	0524 4604 5732 2707	0262 I6I9 6659 4702	I4
I5	2I70 8634 7046 I99I	0000 0000 0000 0000	0275 4787 9I02 3427	I5
I6	0,2452 4926 I076 9962	0,0574 9882 0III 2057	0,0287 55I3 0I49 7085	I6
I7	2745 4950 9537 6309	0000 0000 0000 0000	0298 22I3 5626 0I04	I7
I8	3048 4948 0984 8546	06I5 2454 2I53 3648	0307 5686 0079 5338	I8
I9	3360 2I32 7I44 I073	0000 0000 0000 0000	03I5 6495 2225 I407	I9
20	3679 32I5 95I4 8275	0644 5286 I094 04II	0322 3I48 4087 3956	20
2I	0,4004 3266 4503 2535	0,0000 0000 0000 0000	0,0327 4402 9765 862I	2I
22	4333 7I58 7850 7669	0662 3I0I 9702 3483	033I I065 2377 07I4	22
23	4666 08I7 I296 687I	0000 0000 0000 0000	0333 3904 72I7 8897	23
24	0,5000 0000 0000 0000	0,0668 2728 6093 053I	0,0334 I843 57II 0530	24

No.	Nodes		Weights of Gaussian quadratures		Weights of improved quadratures		No.
	Mantissa	Order	Mantissa	Order	Mantissa	Order	
I	7074 2I6I 7I32 682I	-I3	0000 0000 0000 0000	0	4625 5202 2770 46I4	-II	I
2	5266 2777 437I 424I	-I0	6673 6552 3600 6564	- 7	6550 34II 405I 7767	- I0	2
3	7I53 36I6 4I37 2674	- 7	0000 0000 0000 0000	0	5552 0724 52I3 III0	- 7	3
4	70I6 7234 4II3 2645	- 6	7735 52I6 7562 5336	- 6	7757 6563 6476 7205	- 7	4
5	56I6 4425 I547 7725	- 5	0000 0000 0000 0000	0	5042 0776 I776 4I72	- 6	5
6	4224 3I64 3373 7735	- 4	6II4 I446 24I7 4723	- 5	6I06 7054 2340 II60	- 6	6
7	5752 3455 55I4 I570	- 4	0000 0000 0000 0000	0	7I47 5030 0II0 0773	- 6	7
I0	7706 4502 3504 6365	- 4	4070 6III 5245 34I6	- 4	4072 4362 0573 3305	- 5	I0
II	502I 3223 0276 2552	- 3	0000 0000 0000 0000	0	4464 3464 6330 5777	- 5	II
I2	6I74 5525 5607 4057	- 3	5045 7II2 754I 3777	- 4	5044 4850 7454 036I	- 5	I2
I3	7442 7525 0603 I5I5	- 3	0000 0000 0000 0000	0	54I3 533I 4245 5206	- 5	I3
I4	4400 5636 I2I6 3044	- 2	5744 5237 0I0I 27II	- 4	5745 565I I024 4623	- 5	I4
I5	52I2 I777 5305 2445	- 2	0000 0000 0000 0000	0	6260 0526 6376 6582	- 5	I5
I6	6054 II76 4852 I0I6	- 2	6555 0652 I47I 40I6	- 4	6554 I58I 32I6 28I3	- 5	I6
I7	6744 5742 I076 I33I	- 2	0000 0000 0000 0000	0	7032 6026 7643 I874	- 5	I7
20	7662 I237 2354 3204	- 2	7270 I533 6I63 6667	- 4	7270 7740 7453 5042	- 5	20
2I	43II 0700 6623 0645	- I	0000 0000 0000 0000	0	7504 66I5 II53 645I	- 5	2I
22	470I 2473 I34I 0247	- I	7700 0450 4I63 I60I	- 4	7677 27I7 4I26 5I65	- 5	22
23	5300 5374 6I32 7526	- I	0000 0000 0000 0000	0	4024 5I00 260I 5707	- 4	23
24	5706 0632 50I0 3222	- I	4077 7734 4504 I242	- 3	4I00 245I I0I6 4430	- 4	24
25	6320 2602 47II 25I6	- I	0000 0000 0000 0000	0	4I4I 7232 4373 636I	- 4	25
26	6736 I34I 3632 5525	- I	4I72 2040 77I5 06I3	- 3	4I7I 74II 0433 6554	- 4	26
27	7856 3504 II70 400I	- I	0000 0000 0000 0000	0	42I0 7206 2573 0257	- 4	27
30	4000 0000 0000 0000	0	42I5 6276 4040 0063	- 3	42I6 0705 0I66 3063	- 4	30

No.	Nodes	Weights of Gaussian quadratures	Weights of improved quadratures	No.
25	0,5333 9I82 8703 3I29	0,0000 0000 0000 0000	0,0333 3904 72I7 8897	25
26	5666 284I 2I49 233I	0662 3I0I 9702 3483	033I I065 2377 07I4	26
27	5995 6733 5496 7465	0000 0000 0000 0000	0327 4402 9765 862I	27
28	0,6320 6784 0485 I725	0,0644 5286 I094 04II	0,0322 3I48 4087 3956	28
29	6639 7867 2855 8927	0000 0000 0000 0000	03I5 6495 2225 I407	29
30	695I 505I 90I5 I454	06I5 2454 2I53 3648	0307 5686 0079 5338	30
3I	7254 5049 0462 369I	0000 0000 0000 0000	0298 22I3 5626 0I04	3I
32	7547 5073 8923 0038	0574 9832 0III 2057	0287 55I3 0I49 7085	32
33	0,7829 I365 2953 8009	0,0000 0000 0000 0000	0,0275 4787 9I02 3427	33
34	8098 0493 788I 823I	0524 4604 5732 2707	0262 I6I9 6659 4702	34
35	8353 II89 45I8 52I2	0000 0000 0000 0000	0247 809I I975 0I02	35
36	8593 3068 I565 975I	0464 5788 3C30 0I76	0232 3706 6585 6244	36
37	88I7 4565 8II0 5827	0000 0000 0000 0000	02I5 7347 I9I2 0782	37
38	0,9024 4420 0809 4I99	0,0396 4070 5888 3595	0,0I98 I0I5 6800 3308	38
39	92I3 4866 I498 5065	0000 0000 0000 0000	0I79 7828 9728 8304	39
40	9383 76I7 9I35 2208	032I I62I 0704 2629	0I60 7I77 05I2 7460	40
4I	9534 5384 7370 6407	0000 0000 0000 0000	0I40 67I9 5372 29I3	4I
42	9664 8554 34I3 008I	0240 I883 5865 5423	0II9 89I5 7930 69I4	42
43	0,9774 279I 9930 0556	0,0000 0000 0000 0000	0,0098 9579 99I4 0407	43
44	9862 7I23 5609 0576	0I54 9400 2928 4897	0077 8I67 8730 5560	44
45	9929 5907 5363 6878	0000 0000 0000 0000	0055 7349 094I 4I76	45
46	9973 8466 7498 776I	0067 0592 9743 5709	0032 7342 9700 698I	46
47	9995 6556 823I 7I07	0000 0000 0000 0000	00II 6999 4749 7333	47
Remainder terms		$G_{46} = 0{,}I04 \times I0^{-II}$	$K_{72} = 0{,}394 \times I0^{-22}$	

$$n = 27$$

No.	Nodes Mantissa	Order	Weights of Gaussian quadratures Mantissa	Order	Weights of improved quadratures Mantissa	Order	No.
3I	42I0 6I35 7303 5777	0	0000 0000 0000 0000	0	42I0 7206 2573 0257	− 4	3I
32	4420 72I7 2062 5I25	0	4I72 2040 77I5 06I3	− 3	4I7I 74II 0433 6554	− 4	32
33	4627 6476 5433 2530	0	0000 0000 0000 0000	0	4I4I 7232 4373 636I	− 4	33
34	5034 7462 5373 6266	0	4077 7734 4504 I242	− 3	4I00 245I I0I6 4430	− 4	34
35	5237 520I 4722 4I24	0	0000 0000 0000 0000	0	4024 5I00 260I 5707	− 4	35
36	5437 2542 32I7 3654	0	7700 0450 4I63 I60I	− 4	7677 27I7 4I26 5I65	− 5	36
37	5633 3437 4466 3455	0	0000 0000 0000 0000	0	7504 66I5 II53 645I	− 5	37
40	6023 3530 I304 7I36	0	7270 I533 6I63 6667	− 4	7270 7740 7453 5042	− 5	40
4I	6206 6407 3560 35II	0	0000 0000 0000 0000	0	7032 6026 7643 I374	− 5	4I
42	6364 7540 2705 3574	0	6555 0652 I47I 40I6	− 4	6554 I53I 32I6 28I3	− 5	42
43	6535 3400 05I6 5266	0	0000 0000 0000 0000	0	6260 0526 6376 6532	− 5	43
44	6677 6430 3534 3I66	0	5744 5237 0I0I 27II	− 4	5745 565I I024 4623	− 5	44
45	7033 5025 27I7 4626	0	0000 0000 0000 0000	0	54I3 533I 4245 5206	− 5	45
46	7I60 3225 22I7 0372	0	5045 7II2 754I 3777	− 4	5044 4850 7454 036I	− 5	46
47	7275 6455 4750 I522	0	0000 0000 0000 0000	0	4464 3464 6330 5777	− 5	47
50	7403 4553 66I3 5460	0	4070 6III 5245 34I6	− 4	4072 4362 0573 3305	− 5	50
5I	750I 26I5 III3 I7I0	0	0000 0000 0000 0000	0	7I47 5030 0II0 0773	− 6	5I
52	7566 5630 5620 2002	0	6II4 I446 24I7 4723	− 5	6I06 7054 2340 II60	− 6	52
53	7643 4267 2544 600I	0	0000 0000 0000 0000	0	5042 0776 I776 4I72	− 6	53
54	7707 6I05 4336 645I	0	7735 52I6 7562 5336	− 6	7757 6563 6476 7205	− 7	54
55	7743 I220 7057 2024	0	0000 0000 0000 0000	0	5552 0724 52I3 III0	− 7	55
56	7765 2232 0007 0I47	0	6673 6552 3600 6564	− 7	6550 34II 405I 7767	−I0	56
57	7776 I607 3434 I5I2	0	0000 0000 0000 0000	0	4625 5202 2770 46I4	−II	57

No.	Nodes	Weights of Gaussian quadratures	Weights of improved quadratures	No.
I	0,0003 9947 I989 0625	0,0000 0000 0000 0000	0,00I0 76I5 4275 473I	I
2	0024 0639 000I 4893	006I 706I 4899 9936	0030 I283 5507 860I	2
3	0064 7975 I992 0955	0000 0000 0000 0000	005I 2989 3204 6404	3
4	0I26 3572 20I4 3453	0I42 6569 43I4 4668	007I 6372 23I5 44I9	4
5	0207 79I5 7797 3953	0000 0000 0000 0000	009I I558 9275 I936	5
6	0,0308 6272 8998 6336	0,022I 387I 9408 7099	0,0II0 5204 2450 0309	6
7	0428 7965 4602 5442	0000 0000 0000 0000	0I29 7559 7330 687I	7
8	0567 9223 6497 7995	0296 4929 2457 7I84	0I48 3565 3045 3296	8
9	0725 2309 8047 9743	0000 0000 0000 0000	0I66 I368 9I59 9I48	9
I0	0899 9900 70I3 0485	0366 7324 0705 5402	0I83 2905 0I2I I065	I0
II	0,I09I 6I38 676I 7678	0,0000 0000 0000 0000	0,0I99 888I I946 8III	II
I2	I299 3790 42I0 7228	0480 9508 0765 9766	02I5 5296 4096 8476	I2
I3	I522 3396 380I 6I06	0000 0000 0000 0000	0230 2294 5888 28I9	I3
I4	I759 53I7 403I 5I22	0488 0932 6052 0569	0244 0069 3376 6295	I4
I5	20I0 0474 3046 9608	0000 0000 0000 0000	0256 8597 569I 7288	I5
I6	0,2272 8926 4305 5802	0,0537 22I3 5057 9828	0,0268 6397 2750 8757	I6
I7	2546 938I 7268 2768	0000 0000 0000 0000	0279 2585 755I 53I7	I7
I8	283I 0324 6I86 9774	0577 5283 4026 8628	0288 7433 7687 8450	I8
I9	3I24 0442 260I 0246	0000 0000 0000 0000	0297 0827 I622 9760	I9
20	3424 7866 0I5I 9I83	0608 3523 6463 90I7	0304 I90I 7436 5639	20
2I	0,373I 9978 48I5 I6I0	0,0000 0000 0000 0000	0,03I0 0200 0709 8906	2I
22	4044 4056 6263 I9I8	0629 I872 8I73 4I4I	03I4 5855 6I34 849I	22
23	4360 7437 9856 89I6	0000 0000 0000 0000	03I7 8743 9356 4862	23
24	4679 7I55 3568 6972	0639 6909 7673 376I	03I9 848I 3I20 6882	24
25	0,5000 0000 0000 0000	0,0000 0000 0000 0000	0,0320 5023 I884 6334	25

No.	Nodes Mantissa	Order	Weights of Gaussian quadratures Mantissa	Order	Weights of improved quadratures Mantissa	Order	No.
I	6427 007I 0I70 7582	−I3	0000 0000 0000 0000	0	4320 6676 6545 I673	−II	I
2	4733 2206 267I 4766	−I0	6243 I336 46I0 4420	− 7	6I27 I375 7627 3763	−I0	2
3	6505 203I 3443 2363	− 7	0000 0000 0000 0000	0	520I 425I 6I36 0046	− 7	3
4	6360 30I7 I500 36II	− 6	7233 5250 5044 I570	− 6	7253 6650 4I77 7II7	− 7	4
5	5243 44I5 3I33 2034	− 5	0000 0000 0000 0000	0	4525 46I5 35I0 3302	− 6	5
6	77I5 I722 6I72 533I	− 5	5525 6I02 3653 7II3	− 5	552I I640 I60I 735I	− 6	6
7	5372 I223 5056 4I0I	− 4	0000 0000 0000 0000	0	65II 363I 427I 5577	− 6	7
I0	72II 737I 7I54 5702	− 4	7456 I422 535I I366	− 5	746I 0475 I223 7540	‒ 6	I0
II	45I0 3375 345I 026I	− 3	0000 0000 0000 0000	0	420I 4556 3467 I242	− 5	II
I2	5605 0546 I745 4440	− 3	4543 3256 0567 7603	− 4	4542 33I5 6654 3734	− 5	I2
I3	677I 000I 2420 2674	− 3	0000 0000 0000 0000	0	5073 2427 2605 4730	− 5	I3
I4	4I20 7I6I II06 I042	− 2	54I0 2I67 5202 I06I	− 4	54I0 7727 3733 027I	− 5	I4
I5	4676 I470 2767 4464	− 2	0000 0000 0000 0000	0	57II 5235 75I3 5I33	− 5	I5
I6	5502 642I 5005 7367	− 2	6I76 6III 5442 II30	− 4	6I76 I766 4032 6324	− 5	I6
I7	6335 2057 7556 44I2	− 2	0000 0000 0000 0000	0	6446 5545 I437 5I73	− 5	I7
20	72I3 7204 2473 I2I5	− 2	6700 5675 6445 I506	− 4	670I 0725 43I7 54I5	− 5	20
2I	4046 3472 2765 II00	− I	0000 0000 0000 0000	0	7II4 2304 4I73 08I7	− 5	2I
22	44I7 I350 4770 3555	− I	73I0 7074 2545 433I	− 4	73I0 4740 0040 I67I	− 5	22
23	4777 I570 7506 2362	− I	0000 0000 0000 0000	0	7465 7302 7025 I340	− 5	23
24	5365 4534 7235 6I33	− I	7622 7I36 34I6 4377	− 4	7623 05I5 5042 0273	− 5	24
25	576I 20I2 6324 76I2	− I	0000 0000 0000 0000	0	7737 3750 3I2I 5406	− 5	25
26	636I I325 3650 6540	− I	40I5 56I0 5747 6I36	− 3	40I5 5257 6707 2370	− 4	26
27	6764 2444 I003 3670	− I	0000 0000 0000 0000	0	4043 I6I3 6I65 5345	− 4	27
30	737I 4767 530I II35	− I	4060 I072 7473 067I	− 3	4060 I20I 7I22 3406	− 4	30
3I	4000 0000 0000 0000	0	0000 0000 0000 0000	0	4064 3432 4706 0444	− 4	3I

No.	Nodes	Weights of Gaussian quadratures	Weights of improved quadratures	No.
26	0,5320 2844 6431 3028	0,0639 6909 7673 3761	0,0319 8481 3120 6882	26
27	5639 2562 0143 1083	0000 0000 0000 0000	0317 8743 9356 4862	27
28	5955 5943 3736 8082	0629 1872 8173 4141	0314 5855 6134 8491	28
29	6268 0021 5184 8390	0000 0000 0000 0000	0310 0200 0709 8906	29
30	0,6575 2133 9848 0817	0,0608 8523 6463 9017	0,0304 1901 7436 5639	30
31	6875 9557 7398 9754	0000 0000 0000 0000	0297 0827 1622 9760	31
32	7168 9675 3813 0226	0577 5283 4026 8628	0288 7433 7687 8450	32
33	7453 0618 2731 7232	0000 0000 0000 0000	0279 2585 7551 5817	33
34	7727 1073 5694 4198	0537 2213 5057 9828	0268 6397 2750 8757	34
35	0,7989 9525 6953 0392	0,0000 0000 0000 0000	0,0256 8597 5691 7288	35
36	8240 4682 5968 4878	0488 0932 6052 0569	0244 0069 3376 6295	36
37	8477 6603 6198 3894	0000 0000 0000 0000	0230 2294 5888 2819	37
38	8700 6209 5789 2772	0430 9508 0765 9766	0215 5296 4096 8476	38
39	8908 3861 3238 2322	0000 0000 0000 0000	0199 8881 1946 8111	39
40	0,9100 0099 2986 9515	0,0366 7324 0705 5402	0,0183 2905 0121 1065	40
41	9274 7690 1952 0257	0000 0000 0000 0000	0166 1368 9159 9148	41
42	9432 0776 3502 2005	0296 4929 2457 7184	0148 8565 3045 3296	42
43	9571 2034 5397 4558	0000 0000 0000 0000	0129 7559 7330 6871	43
44	9691 3727 6001 3664	0221 3871 9408 7099	0110 5204 2450 0309	44
45	0,9792 2084 2202 6047	0,0000 0000 0000 0000	0,0091 1558 9275 1936	45
46	9873 6427 7985 6547	0142 6569 4314 4668	0071 6372 2315 4419	46
47	9935 2024 8007 9045	0000 0000 0000 0000	0051 2989 3204 6404	47
48	9975 9360 9998 5107	0061 7061 4899 9936	0030 1283 5507 8601	48
49	9996 0052 8010 9375	0000 0000 0000 0000	0010 7615 4275 4731	49
Remainder terms		$G_{48} = 0,271 \times 10^{-12}$	$K_{74} = 0,453 \times 10^{-23}$	

No.	Nodes Mantissa	Order	Weights of Gaussian quadratures Mantissa	Order	Weights of improved quadratures Mantissa	Order	No.
32	4203 I404 I287 332I	0	4060 I072 7473 067I	− 3	4060 I20I 7I22 3406	− 4	32
33	4405 6555 7376 2043	0	0000 0000 0000 0000	0	4043 I6I3 6I65 5345	− 4	33
34	4607 3225 2053 45I7	0	40I5 56I0 5747 6I36	− 3	40I5 5257 6707 2370	− 4	34
35	5007 2772 4625 4072	0	0000 0000 0000 0000	0	7737 3750 3I2I 5406	− 5	35
36	5205 I52I 426I 0722	0	7622 7I36 34I6 4377	− 4	7623 05I5 5042 0273	− 5	36
37	5400 3I03 4I34 6606	0	0000 0000 0000 0000	0	7465 7302 7025 I340	− 5	37
40	5570 32I3 5403 6III	0	73I0 7074 2545 433I	− 4	73I0 4740 0040 I67I	− 5	40
4I	5754 6I42 6405 3337	0	0000 0000 0000 0000	0	7II4 2304 47I3 03I7	− 5	4I
42	6I35 0I36 726I I534	0	6700 5675 6445 I506	− 4	670I 0725 43I7 54I5	− 5	42
43	63I0 5364 0044 2675	0	0000 0000 0000 0000	0	6446 5545 I437 5I73	− 5	43
44	6457 2273 4576 4I02	0	6I76 6III 5442 II30	− 4	6I76 I766 4032 6324	− 5	44
45	6620 346I 7202 0662	0	0000 0000 0000 0000	0	57II 5235 75I3 5I33	− 5	45
46	6753 6I43 5556 3567	0	54I0 2I67 5202 I06I	− 4	54I0 7727 3733 027I	− 5	46
47	7I00 6777 6535 75I0	0	0000 0000 0000 0000	0	5073 2427 2605 4730	− 5	47
50	72I7 2723 I603 2333	0	4543 3256 0567 7603	− 4	4542 33I5 6654 3734	− 5	50
5I	7326 7440 2432 675I	0	0000 0000 0000 0000	0	420I 4556 3467 I242	− 5	5I
52	7427 3020 303I I503	0	7456 I422 535I I366	− 5	746I 0475 I223 7540	− 6	52
53	7520 2726 6I35 0573	0	0000 0000 0000 0000	0	65II 363I 427I 5577	− 6	53
54	760I 454I 3234 I25I	0	5525 6I02 3653 7II3	− 5	552I I640 I60I 735I	− 6	54
55	7652 7067 45I5 II37	0	0000 0000 0000 0000	0	4525 46I5 35I0 3302	− 6	55
56	77I4 I747 6062 774I	0	7233 5250 5044 I570	− 6	7253 6650 4I77 7II7	− 7	56
57	7745 3527 632I 5626	0	0000 0000 0000 0000	0	520I 425I 6I36 0046	− 7	57
60	7766 III3 3632 2I46	0	6243 I336 46I0 4420	− 7	6I27 I375 7627 3763	−I0	60
6I	7776 272I 76I5 74I6	0	0000 0000 0000 0000	0	4320 6676 6545 I673	−II	6I

No.	Nodes	Weights of Gaussian quadratures	Weights of improved quadratures	No.
I	0,0008 6894 7503 6951	0,0000 0000 0000 0000	0,0009 9369 1946 1652	I
2	0022 2151 5104 7510	0056 9689 9250 5131	0027 8096 6067 6784	2
3	0059 8210 2732 9614	0000 0000 0000 0000	0047 3698 6693 0871	3
4	0116 6803 9270 2412	0131 7749 3307 5161	0066 1811 4597 7858	4
5	0191 9250 6787 0787	0000 0000 0000 0000	0084 2390 8854 5641	5
6	0,0285 1271 4385 5128	0,0204 6957 8850 6532	0,0102 1768 5572 9414	6
7	0396 2644 2359 1492	0000 0000 0000 0000	0120 0497 2803 4766	7
8	0525 0400 1060 8623	0274 5234 7987 9176	0137 3765 8793 9259	8
9	0670 7646 7353 3622	0000 0000 0000 0000	0153 9615 0083 6937	9
10	0832 7868 5619 5830	0340 1916 6906 1785	0170 0106 5137 1647	10
II	0,1010 6310 1000 7500	0,0000 0000 0000 0000	0,0185 5813 5741 7078	II
12	1203 7036 8481 3212	0400 7035 0167 5005	0200 4191 2752 0162	12
13	1411 1679 6593 4578	0000 0000 0000 0000	0214 3642 2510 0850	13
14	1632 1681 5763 2658	0455 1413 0991 4818	0227 5145 6524 9609	14
15	1865 9495 0494 8413	0000 0000 0000 0000	0239 9126 8569 4184	15
16	0,2111 6853 4879 3885	0,0502 6797 4533 5253	0,0251 3883 9540 3578	16
17	2368 3735 7832 6404	0000 0000 0000 0000	0261 8144 2903 2037	17
18	2634 9863 4277 1425	0542 5981 2287 1318	0271 2556 4944 2727	18
19	2910 5730 8903 4811	0000 0000 0000 0000	0279 7540 5610 2062	19
20	3194 1384 7095 3061	0574 2912 9572 8558	0287 1855 8180 7839	20
21	0,3484 5523 0534 4461	0,0000 0000 0000 0000	0,0293 4484 0011 1971	21
22	3780 6655 8139 5058	0597 2788 1767 8924	0298 6017 0162 0870	22
23	4081 4053 0289 4756	0000 0000 0000 0000	0302 6972 7688 0229	23
24	4385 6765 3694 6448	0611 2122 1495 1550	0305 6425 4858 5265	24
25	4692 2775 8497 1575	0000 0000 0000 0000	0307 3559 4935 7127	25
26	0,5000 0000 0000 0000	0,0615 8802 6863 3577	0,0307 9040 9033 9165	26

$$n = 31$$

No.	Nodes Mantissa	Order	Weights of Gaussian quadratures Mantissa	Order	Weights of improved quadratures Mantissa	Order	No.
I	6026 75I3 5455 007I	-I3	0000 0000 0000 0000	0	4043 7304 6447 43I6	-II	I
2	443I 3326 56I7 2044	-I0	5652 64I5 770I 5222	- 7	5544 0336 2375 6343	-I0	2
3	6I00 2603 63I3 4207	- 7	0000 0000 0000 0000	0	4663 427I 3555 2657	- 7	3
4	5762 55I5 6II3 72I3	- 6	6576 3I5I 5462 I572	- 6	66I5 6304 702I 6034	- 7	4
5	4723 4632 4765 I003	- 5	0000 0000 0000 0000	0	4240 2I57 2043 53I0	- 6	5
6	723I I576 7634 5I7I	- 5	5I72 772I I450 334I	- 5	5I66 4024 260I 5037	- 6	6
7	5044 7526 0775 2I72	- 4	0000 0000 0000 0000	0	6II3 020I 3I22 I650	- 6	7
I0	6560 7I57 3567 I347	- 4	70I6 I677 0634 7375	- 5	702I I752 637I 27I6	- 6	I0
II	4225 7543 03I6 4505	- 3	0000 0000 0000 0000	0	7704 0042 2I40 0I35	- 6	II
I2	5250 7003 7630 242I	- 3	4265 3656 45I4 46I0	- 4	4264 272I 2605 6I70	- 5	I2
I3	6357 5053 6320 5546	- 3	0000 0000 0000 0000	0	.4600 3473 2053 55I4	- 5	I3
I4	7550 2275 3004 3447	- 3	5I02 03I6 5602 I620	- 4	5I02 7357 7643 8380	- 5	I4
I5	44I0 0353 7222 2007	- 2	0000 0000 0000 0000	0	537I 5557 5477 4076	- 5	I5
I6	5I62 III7 0567 0442	- 2	5646 6406 4035 202I	- 4	5646 0503 I554 0433	- 5	I6
I7	576I I277 II76 2565	- 2	0000 0000 0000 0000	0	6II0 4526 I7I2 2604	- 5	I7
20	6603 6220 3720 0474	- 2	6336 27I2 5250 2676	- 4	6336 7767 5536 6544	- 5	20
2I	7450 2576 0227 I357	- 2	0000 0000 0000 0000	0	6547 5I67 I020 0324	- 5	2I
22	4I56 45I3 0030 2640	- I	6743 76II 34I3 5263	- 4	6743 3I56 6244 342I	- 5	22
23	4520 2566 5340 0443	- I	0000 0000 0000 0000	0	7I22 6255 4I40 2575	- 5	23
24	5070 5066 I545 I750	- I	7263 53I6 4582 32I5	- 4	7264 I456 4I3I 3570	- 5	24
25	5446 427I 2550 3772	- I	0000 0000 0000 0000	0	7406 2227 0II7 0464	- 5	25
26	603I 0760 4735 3506	- I	75I2 247I I332 I645	- 4	75II 6520 623I I265	- 5	26
27	64I7 37I3 5270 374I	- I	0000 0000 0000 0000	0	7577 4070 2365 I404	- 5	27
30	70I0 5760 4066 3546	- I	7645 5076 755I 627I	- 4	7646 0743 3026 4III	- 5	30
3I	7403 7286 7243 67II	- I	0000 0000 0000 0000	0	7674 4466 7044 6674	- 5	3I
32	4000 0000 0000 0000	0	7704 I672 0464 0255	- 4	7703 6052 765I 4246	- 5	32

No.	Nodes	Weights of Gaussian quadratures	Weights of improved quadratures	No.
27	0,5307 7224 I502 8425	0,0000 0000 0000 0000	0,0307 3559 4935 7I27	27
28	56I4 3234 6305 3552	06II 2I22 I495 I550	0305 6425 4858 5265	28
29	59I8 5946 97I0 5244	0000 0000 0000 0000	0302 6972 7688 0229	29
30	62I9 8344 I860 4942	0597 2788 I767 8924	0298 60I7 0I62 0870	30
3I	65I5 4476 9465 5539	0000 0000 0000 0000	0293 4484 00II I97I	3I
32	0,6805 86I5 2904 6939	0,0574 29I2 9572 8558	0,0287 I855 8I80 7839	32
33	7089 4269 I096 5I89	0000 0000 0000 0000	0279 7540 56I0 2062	33
34	7365 0I36 5722 8575	0542 598I 2237 I3I8	027I 2556 4944 2727	34
35	763I 6264 2I67 3596	0000 0000 0000 0000	026I 8I44 2903 2037	35
36	7888 3I46 5I20 6II5	0502 6797 4533 5253	025I 3883 9540 3578	36
37	0,8I34 0504 9505 I587	0,0000 0000 0000 0000	0,0239 9I26 8569 4I84	37
38	8367 83I8 4236 7342	0455 I4I3 099I 48I8	0227 5I45 6524 9609	38
39	8588 8320 3406 5422	0000 0000 0000 0000	02I4 3642 25I0 0850	39
40	8796 2963 I5I8 6788	0400 7035 0I67 5005	0200 4I9I 2752 0I62	40
4I	8989 3689 8999 2500	0000 0000 0000 0000	0I85 58I3 574I 7078	4I
42	0,9I67 2I3I 4380 4I70	0,0340 I9I6 6906 I785	0,0I70 0I06 5I37 I647	42
43	9329 2353 2646 6378	0000 0000 0000 0000	0I53 96I5 0083 6937	43
44	9474 9599 8939 I377	0274 5234 7987 9I76	0I37 3765 8793 9259	44
45	9603 7355 7640 8508	0000 0000 0000 0000	0I20 0497 2803 4766	45
46	97I4 8728 56I4 4872	0204 6957 8350 6532	0I02 I768 5572 94I4	46
47	0,9808 0749 32I2 92I3	0,0000 0000 0000 0000	0,0084 2390 8854 564I	47
48	9883 3I96 0729 7588	0I3I 7749 3307 5I6I	0066 I8II 4597 7858	48
49	9940 I789 7267 0386	0000 0000 0000 0000	0047 3698 6698 087I	49
50	9977 7848 4895 2490	0056 9689 9250 5I3I	0027 8096 6067 6784	50
5I	9996 3I05 2496 3049	0000 0000 0000 0000	0009 9369 I946 I652	5I
Remainder terms		$G_{50} = 0{,}705 \times 10^{-13}$	$K_{78} = 0{,}543 \times 10^{-24}$	

No.	Nodes Mantissa	Order	Weights of Gaussian quadratures Mantissa	Order	Weights of improved quadratures Mantissa	Order	No.
33	4I76 0260 4256 0433	0	0000 0000 0000 0000	0	7674 4466 7044 6674	− 5	33
34	4373 5007 5744 6II4	0	7645 5076 755I 627I	− 4	7646 0743 3026 4III	− 5	34
35	4570 2082 I243 60I7	0	0000 0000 0000 0000	0	7577 4070 2365 I404	− 5	35
36	4763 3407 542I 2I34	0	75I2 247I I332 I645	− 4	75II 6520 623I I265	− 5	36
37	5I54 5648 25I3 6002	0	0000 0000 0000 0000	0	7406 2227 0II7 0464	− 5	37
40	5343 5344 7II5 30I3	0	7263 53I6 4532 32I5	− 4	7264 I456 4I3I 3570	− 5	40
4I	5527 6504 52I7 7556	0	0000 0000 0000 0000	0	7I22 6255 4I40 2575	− 5	4I
42	57I0 5532 3763 6457	0	6743 76II 34I3 5263	− 4	6743 3I56 6244 342I	− 5	42
43	6065 7240 3732 I504	0	0000 0000 0000 0000	0	6547 5I67 I020 0324	− 5	43
44	6237 0333 70I3 7660	0	6336 27I2 5250 2676	− 4	6336 7767 5536 6544	− 5	44
45	6403 5520 I540 3242	0	0000 0000 0000 0000	0	6II0 4526 I7I2 2604	− 5	45
46	6543 3554 I642 I667	0	5646 6406 4035 202I	− 4	5646 0503 I554 0433	− 5	46
47	6675 7705 0I33 3376	0	0000 0000 0000 0000	0	537I 5557 5477 4076	− 5	47
50	7022 7550 2477 3433	0	5I02 03I6 5602 I620	− 4	5I02 7357 7643 3330	− 5	50
5I	7I42 0272 4I45 7223	0	0000 0000 0000 0000	0	4600 3473 2053 55I4	− 5	5I
52	7252 7077 40I4 7535	0	4265 3656 45I4 46I0	− 4	4264 272I 2605 6I70	− 5	52
53	7355 2023 4746 I327	0	0000 0000 0000 0000	0	7704 0042 2I40 0I35	− 6	53
54	7450 748I 02I0 432I	0	70I6 I677 0634 7375	− 5	702I I752 637I 27I6	− 6	54
55	7535 54I2 4740 I270	0	0000 0000 0000 0000	0	6II3 020I 3I22 I650	− 6	55
56	76I3 I544 0203 0654	0	5I72 772I I450 334I	− 5	5I66 4024 260I 5037	− 6	56
57	766I 3063 I260 2557	0	0000 0000 0000 0000	0	4240 2I57 2043 53I0	− 6	57
60	7720 I522 62I6 6405	0	6576 3I5I 5462 I572	− 6	66I5 6304 702I 6034	− 7	60
6I	7747 3764 7606 32I6	0	0000 0000 0000 0000	0	4663 427I 3555 2657	− 7	6I
62	7766 7I5I I224 34I3	0	5652 64I5 770I 5222	− 7	5544 0336 2375 6343	−I0	62
63	7776 3722 0550 4645	0	0000 0000 0000 0000	0	4043 7304 6447 48I6	−II	63

No.	Nodes	Weights of Gaussian quadratures	Weights of improved quadratures	No.
I	0,0003 4146 7619 9641	0,0000 0000 0000 0000	0,0009 1990 5925 7304	I
2	0020 5714 9427 1915	0052 7568 6308 6715	0025 7586 7973 0094	2
3	0055 4042 7587 7825	0000 0000 0000 0000	0043 8763 8353 6499	3
4	0108 0727 7021 7645	0122 0892 5546 3160	0061 3090 7931 6776	4
5	0177 7944 0880 3437	0000 0000 0000 0000	0078 0762 0464 2687	5
6	0,0264 2046 6669 1429	0,0189 8119 1647 1814	0,0094 7571 8572 5455	6
7	0367 2971 4614 8194	0000 0000 0000 0000	0111 3848 6798 6541	7
8	0486 8106 9007 8465	0254 8791 2648 5739	0127 5345 7561 7041	8
9	0622 1518 2418 1647	0000 0000 0000 0000	0143 0544 1628 6236	9
IO	0772 7702 8605 7510	0316 3702 3164 7874	0158 1192 9595 7622	IO
II	0,0938 2513 6457 4358	0,0000 0000 0000 0000	0,0172 7568 4389 2905	II
I2	1118 0702 5589 6606	0373 4207 4882 8299	0186 7580 3771 6842	I2
I3	1311 5151 0246 1987	0000 0000 0000 0000	0200 0099 7094 8005	I3
I4	1517 8636 9790 0214	0425 2294 7156 7426	0212 5792 8569 4541	I4
I5	1736 4499 5246 5433	0000 0000 0000 0000	0224 4716 4387 6973	I5
I6	0,1966 5385 3491 1910	0,0471 0690 0177 9571	0,0235 5610 9995 8281	I6
I7	2207 2773 4509 5922	0000 0000 0000 0000	0245 7706 9972 9340	I7
I8	2457 7964 2587 7471	0510 2958 0547 2127	0255 1280 6853 3447	I8
I9	2717 2489 1985 8502	0000 0000 0000 0000	0263 6284 8155 4362	I9
20	2984 7412 2438 2568	0542 3592 0264 2883	0271 1939 7768 4047	20
2I	0,3259 3090 1448 1525	0,0000 0000 0000 0000	0,0277 7784 6222 II25	2I
22	3539 9758 0257 0216	0566 8090 8273 1598	0283 3947 7887 6815	22
23	3825 7737 2423 7712	0000 0000 0000 0000	0288 0349 8913 8572	23
24	4115 7058 9821 5549	0583 3022 1742 6483	0291 6567 7676 8922	24
25	4408 7416 8553 9714	0000 0000 0000 0000	0294 2428 8465 3297	25
26	0,4703 8495 3285 3434	0,0591 6070 7639 6311	0,0295 8016 7904 6173	26
27	5000 0000 0000 0000	0000 0000 0000 0000	0296 3251 4330 0198	27

No.	Nodes Mantissa	Order	Weights of Gaussian quadratures Mantissa	Order	Weights of improved quadratures Mantissa	Order	No.
I	5460 3402 0524 3273	-I3	0000 0000 0000 0000	0	7422 2727 3568 I722	-I2	I
2	4I55 0475 4064 2424	-I0	53I5 7652 0507 0405	- 7	52I4 7744 5477 02I7	-I0	2
3	5530 6I7I 5I34 7423	- 7	0000 0000 0000 0000	0	4374 3055 4742 406I	- 7	3
4	542I 040I 2570 066I	- 6	6200 376I 7703 2333	- 6	62I6 27I0 4I75 502I	- 7	4
5	4482 3060 5076 4547	- 5	0000 0000 0000 0000	0	7775 342I 2I55 6327	- 7	5
6	6606 7674 0223 7665	- 5	4667 7I6I 5272 I734	- 5	4664 00I3 2600 I377	- 6	6
7	4547 0745 54I7 7065	- 4	0000 0000 0000 0000	0	5547 7063 I033 6307	- 6	7
I0	6I66 27I4 7522 5427	- 4	64I4 6006 72II 5422	- 5	64I7 I740 62I5 6I00	- 6	I0
II	7755 2530 67I4 7642	- 4	0000 0000 0000 0000	0	7246 0536 77I4 4437	- 6	II
I2	4744 I553 I5I4 0004	- 3	403I 2722 5725 7I60	- 4	4030 4005 0400 I666	- 5	I2
I3	6002 3544 5I20 2367	- 3	0000 0000 0000 0000	0	4330 2674 3357 5006	- 5	I3
I4	7II7 5424 7457 34I3	- 3	46I7 2000 7I42 7502	- 4	46I7 6777 6332 0363	- 5	I4
I5	4I44 6224 6725 0II3	- 2	0000 0000 0000 0000	0	5075 444I 432I 5433	- 5	I5
I6	4666 6742 66I6 4045	- 2	5342 62I2 5536 IIII	- 4	5342 2433 377I 0574	- 5	I6
I7	5434 7776 2754 5273	- 2	0000 0000 0000 0000	0	5576 I435 4735 5753	- 5	I7
20	6225 7640 5530 I027	- 2	60I7 I452 I656 II33	- 4	60I7 4276 2033 I254	- 5	20
2I	7040 3I63 4I02 37I5	- 2	0000 0000 0000 0000	0	6225 273I 7607 6367	- 5	2I
22	7672 6650 4673 7556	- 2	6420 2I44 5624 I424	- 4	6420 0073 7040 7225	- 5	22
23	426I 7606 3222 6475	- I	0000 0000 0000 0000	0	6577 3346 2520 3077	- 5	23
24	46I5 063I 5I22 3064	- I	6742 3I73 77I5 557I	- 4	6742 4577 640I 2227	- 5	24
25	5I56 0I52 2II7 I604	- I	0000 0000 0000 0000	0	7070 7I35 5047 4452	- 5	25
26	5523 7458 5702 II23	- I	7202 5075 3527 I747	- 4	7202 406I 2555 I55I	- 5	26
27	6076 0456 3327 7732	- I	0000 0000 0000 0000	0	7277 2520 5660 6553	- 5	27
30	6453 454I 2750 7332	- I	7356 5653 5300 6265	- 4	7356 6333 7623 3267	- 5	30
3I	7033 5I02 2362 7I26	- I	0000 0000 0000 0000	0	7420 5464 4525 25I2	- 5	3I
32	74I5 3II3 7334 4066	- I	7445 II77 4205 I662	- 4	7445 I033 5605 3057	- 5	32
33	4000 0000 0000 0000	0	0000 0000 0000 0000	0	7453 7742 7744 I4I3	- 5	33

No.	Nodes	Weights of Gaussian quadratures	Weights of improved quadratures	No.
28	0,5296 I504 67I4 6566	0,059I 6070 7639 63II	0,0295 80I6 7904 6I78	28
29	0,559I 2583 I446 0286	0,0000 0000 0000 0000	0,0294 2428 8465 3297	29
30	5884 294I 0I78 445I	0583 3022 I742 6483	029I 6567 7676 8922	30
3I	6I74 2262 7576 2288	0000 0000 0000 0000	0288 0349 89I3 8572	3I
32	6460 024I 9742 9784	0566 8090 8273 I598	0283 3947 7887 68I5	32
33	6740 6909 855I 8474	0000 0000 0000 0000	0277 7784 6222 II25	33
34	0,70I5 2587 756I 7482	0,0542 3592 0264 2883	0,027I I939 7768 4047	34
35	7282 75I0 80I4 I498	0000 0000 0000 0000	0263 6284 8I55 4362	35
36	7542 2035 74I2 2529	05I0 2958 0547 2I27	0255 I280 6853 3447	36
37	7792 7226 5490 4078	0000 0000 0000 0000	0245 7706 9972 9340	37
38	8033 46I4 6508 8090	047I 0690 0I77 957I	0235 56I0 9995 828I	38
39	0,8263 5500 4753 4567	0,0000 0000 0000 0000	0,0224 47I6 4387 6973	39
40	8482 I363 0209 9786	0425 2294 7I56 7426	02I2 5792 8569 454I	40
4I	8688 4848 9753 8063	0000 0000 0000 0000	0200 0099 7094 8005	4I
42	888I 9297 44I0 3394	0373 4207 4882 8299	0I86 7580 377I 6842	42
43	906I 7486 3542 5642	0000 0000 0000 0000	0I72 7568 4389 2905	43
44	0,9227 2297 I394 2490	0,03I6 3702 3I64 7874	0,0I58 II92 9595 7622	44
45	9377 848I 758I 8358	0000 0000 0000 0000	0I43 0544 I628 6236	45
46	95I3 I893 0992 I535	0254 879I 2648 5739	0I27 5345 756I 704I	46
47	9632 7028 5385 I806	0000 0000 0000 0000	0III 3848 6798 654I	47
48	9735 7953 3330 857I	0I89 8II9 I647 I8I4	0094 757I 8572 5455	48
49	0,9822 2055 9II9 6563	0,0000 0000 0000 0000	0,0078 0762 0464 2687	49
50	989I 9272 2978 2355	0I22 0892 5546 3I60	006I 3090 793I 6776	50
5I	9944 5957 24I2 2I75	0000 0000 0000 0000	0043 8763 8353 6499	5I
52	9979 4285 0572 8085	0052 7568 6808 67I5	0025 7586 7973 0094	52
53	9996 5853 2380 0359	0000 0000 0000 0000	0009 I990 5925 7304	53
Remainder terms		$G_{52}= 0,I83 \times I0^{-I3}$	$K_{80}= 0,628 \times I0^{-25}$	

64

No.	Nodes		Weights of Gaussian quadratures		Weights of improved quadratures		No.
	Mantissa	Order	Mantissa	Order	Mantissa	Order	
34	4I7I 2332 022I 5745	0	7445 II77 4205 I662	− 4	7445 I033 5605 3057	− 5	34
35	4362 I336 6606 4325	0	0000 0000 0000 0000	0	7420 5464 4525 25I2	− 5	35
36	4552 I5I7 24I3 4223	0	7356 5653 5300 6265	− 4	7356 6333 7623 3267	− 5	36
37	4740 7550 6224 0023	0	0000 0000 0000 0000	0	7277 2520 5660 6553	− 5	37
40	5I26 0I52 I036 7326	0	7202 5075 3527 I747	− 4	7202 406I 2555 I55I	− 5	40
4I	53I0 77I2 6730 3076	0	0000 0000 0000 0000	0	7070 7I35 5047 4452	− 5	4I
42	547I 3463 I326 6346	0	6742 3I73 77I5 557I	− 4	6742 4577 640I 2227	− 5	42
43	5647 0074 6266 454I	0	0000 0000 0000 0000	0	6577 3346 2520 3077	− 5	43
44	602I 2225 662I 0044	0	6420 2I44 5624 I424	− 4	6420 0073 7040 7225	− 5	44
45	6I67 7I43 0757 30I5	0	0000 0000 0000 0000	0	6225 273I 7607 6367	− 5	45
46	6332 4027 645I 7572	0	60I7 I452 I656 II33	− 4	60I7 4276 2033 I254	− 5	46
47	6470 6000 3204 652I	0	0000 0000 0000 0000	0	5576 I435 4735 5753	− 5	47
50	6622 2207 2234 2767	0	5342 62I2 5536 IIII	− 4	5342 2433 377I 0574	− 5	50
5I	6746 6332 62I2 5755	0	0000 0000 0000 0000	0	5075 444I 432I 5433	− 5	5I
52	7066 0285 3032 0486	0	46I7 2000 7I42 7502	− 4	46I7 6777 6332 0363	− 5	52
53	7I77 5423 3265 754I	0	0000 0000 0000 0000	0	4330 2674 3357 5006	− 5	53
54	7303 3622 4626 3777	0	403I 2722 5725 7I60	− 4	4030 4005 0400 I666	− 5	54
55	740I I252 3443 I406	0	0000 0000 0000 0000	0	7246 0536 77I4 4437	− 6	55
56	7470 4643 I4I2 65I6	0	64I4 6006 72II 5422	− 5	64I7 I740 62I5 6I00	− 6	56
57	755I 434I 5II7 0035	0	0000 0000 0000 0000	0	5547 7063 I033 6307	− 6	57
60	7623 6202 0773 3002	0	4667 7I6I 5272 I734	− 5	4664 00I3 2600 I377	− 6	60
6I	7667 I3I6 3656 0265	0	0000 0000 0000 0000	0	7775 342I 2I55 6327	− 7	6I
62	7723 5673 7652 077I	0	6200 376I 7703 2333	− 6	62I6 27I0 4I75 502I	− 7	62
63	775I 2347 08I3 2I42	0	0000 0000 0000 0000	0	4374 3055 4742 406I	− 7	63
64	7767 4456 6047 6273	0	58I5 7652 0507 0405	− 7	52I4 7744 5477 02I7	−I0	64
65	7776 4637 0773 6527	0	0000 0000 0000 0000	0	7422 2727 3563 I722	−I2	65

n = 27

No.	Nodes	Weights of Gaussian quadratures	Weights of improved quadratures	No.
I	0,0003 I722 6738 5II4	0,0000 0000 0000 0000	0,0008 5443 0248 I755	I
2	00I9 I036 8555 5057	0048 9949 8025 6472	0023 9I76 8277 6597	2
3	005I 45I9 8849 I5II	0000 0000 0000 0000	0040 7539 2449 5906	3
4	0I00 3826 20I9 2494	0II3 43II 5798 0903	0056 9675 3I72 2434	4
5	0I65 I753 7I86 8784	0000 0000 0000 0000	0072 5644 I387 3858	5
6	0,0245 4972 I092 6475	0,0I76 4852 6878 7099	0,0088 0962 6738 8255	6
7	034I 3703 8772 5059	0000 0000 0000 0000	0I03 6I69 I593 0595	7
8	0452 5883 966I 2544	0237 2470 6260 3075	0II8 72I4 69I0 7I90	8
9	0578 6I67 0I44 9054	0000 0000 0000 0000	0I33 2504 6356 8305	9
IO	07I8 9604 5990 8528	0294 9I76 8429 9I68	0I47 3865 4I22 6756	IO
II	0,0873 2804 6376 8I90	0,0000 0000 0000 0000	0,0I6I I83I 6395 8325	II
I2	I04I I4I8 0464 7459	0348 744I I883 I228	0I74 4289 I382 3230	I2
I3	I22I 9082 4786 3856	0000 0000 0000 0000	0I86 9939 7366 25I7	I3
I4	I4I4 9326 3I30 2882	0398 0243 3886 5289	0I98 965I 7645 37I9	I4
I5	I6I9 6558 6557 7478	0000 0000 0000 0000	02I0 3802 494I 8486	I5
I6	0,I835 460I 4026 7524	0,0442 II57 927I 8785	0,022I 0982 0654 4328	I6
I7	206I 5829 4063 9459	0000 0000 0000 0000	023I 0I5I 6420 70I0	I7
I8	2297 242I 77I0 27I6	0480 4436 3685 0I43	0240 I86I I329 44I5	I8
I9	254I 7I96 9293 9I45	0000 0000 0000 0000	0248 6444 9476 6087	I9
20	2794 2587 4I24 9866	05I2 508I 8908 8729	0256 2865 6024 5634	20
2I	0,3053 9907 3626 3976	0,0000 0000 0000 0000	0,0263 028I 7374 9038	2I
22	3320 0304 8I80 7456	0537 89I4 2894 2666	0268 9I54 70I2 5583	22
23	359I 55I0 9339 9963	0000 0000 0000 0000	0273 9857 7478 636I	23
24	3867 703I 7280 23I6	0556 2624 4I78 4226	0278 I599 990I 754I	24
25	4I47 5478 4574 5624	0000 0000 0000 0000	028I 3696 6805 9756	25
26	0,4430 I370 7I95 2350	0,0567 38I7 3054 4826	0,0283 6629 0926 4I86	26
27	47I4 5854 I638 4063	0000 0000 0000 0000	0285 0865 5073 3220	27
28	5000 0000 0000 0000	057I I043 3689 4785	0285 5798 5063 7827	28

No.	Nodes		Weights of Gaussian quadratures		Weights of improved quadratures		No.
	Mantissa	Order	Mantissa	Order	Mantissa	Order	
I	5I45 0563 572I I774	−I3	0000 0000 0000 0000	0	6777 5727 7077 4I33	−I2	I
2	7646 2524 7403 5573	−II	50I0 5767 7II3 5570	− 7	47I3 7466 4203 56I4	−I0	2
3	52II 44I6 3II6 7I0I	− 7	0000 0000 0000 0000	0	4I30 5336 5I20 4646	− 7	3
4	5I07 3605 6I23 2353	− 6	5635 4I7I 66I4 6557	− 6	5652 5724 22I3 47I2	− 7	4
5	4I64 77II 20I6 3077	− 5	0000 0000 0000 0000	0	7334 356I I6I5 5I23	− 7	5
6	622I 6I77 II60 030I	− 5	44II I644 56I0 4I07	− 5	4405 3I00 5356 6437	− 6	6
7	4275 I507 4434 I535	− 4	0000 0000 0000 0000	0	5234 2025 5027 7733	− 6	7
I0	5626 0525 2063 I456	− 4	6045 5I20 5375 4346	− 5	6050 I544 5334 0007	− 6	I0
II	7320 0I33 6366 67I6	− 4	0000 0000 0000 0000	0	6645 05I3 4507 752I	− 6	II
I2	4463 7073 7437 0I74	− 3	743I 4270 4720 5734	− 5	7427 5I45 2706 464I	− 6	I2
I3	5455 44I3 7633 I06I	− 3	0000 0000 0000 0000	0	4I00 52I7 I622 6I76	− 5	I3
I4	6523 4720 6022 5833	− 3	4355 4I70 520I 2575	− 4	4356 2I45 002I 6273	− 5	I4
I5	7643 7456 7052 0535	− 3	0000 0000 0000 0000	0	4622 7572 4III I440	− 5	I5
I6	44I6 I634 II57 7342	− 2	5060 3740 3730 0I42	− 4	5057 7005 4456 5307	− 5	I6
I7	5I35 5II6 4I25 7467	− 2	0000 0000 0000 0000	0	5805 3757 5062 0040	− 5	I7
20	5677 I574 3743 4547	− 2	552I 3463 3423 6I25	− 4	552I 7647 4503 I350	− 5	20
2I	646I 5450 7263 3426	− 2	0000 0000 0000 0000	0	5723 7544 I407 5064	− 5	2I
22	7263 6323 22I2 430I	− 2	6II4 5052 53I0 6I32	− 4	6II4 I255 6II7 4082	− 5	22
23	4042 I324 0365 5304	− I	0000 0000 0000 0000	0	6273 0207 5253 4366	− 5	23
24	486I 0350 3740 57I2	− I	6436 6I40 7433 2620	− 4	6437 I457 7356 4354	− 5	24
25	4705 6504 3456 4I74	− I	0000 0000 0000 0000	0	6567 440I 4325 7465	− 5	25
26	5237 6II5 5503 3634	− I	6705 I00I 0634 2635	− 4	6704 565I 2752 3053	− 5	26
27	5576 I455 543I 637I	− I	0000 0000 0000 0000	0	7007 I373 2I64 4406	− 5	27
30	6I40 3302 237I 0I64	− I	7075 4I30 I6I3 5854	− 4	7075 7I4I I704 3073	− 5	30
3I	6505 5275 I563 3354	− I	0000 0000 0000 0000	0	7I47 7577 0I73 02I7	− 5	3I
32	7055 I26I 2402 5240	− I	7206 3III 2664 I272	− 4	7206 0I54 3654 7453	− 5	32
33	7426 I403 4470 I574	− I	0000 0000 0000 0000	0	7230 5373 5I42 5755	− 5	33
34	4000 0000 0000 0000	0	7236 624I 2364 7226	− 4	7237 II57 3706 7560	− 5	34

No.	Nodes	Weights of Gaussian quadratures	Weights of improved quadratures	No.
29	0,5285 4145 8361 5937	0,0000 0000 0000 0000	0,0285 0865 5073 3220	29
30	5569 8629 2804 7650	0567 3817 3054 4826	0283 6629 0926 4186	30
31	0,5852 4521 5425 4376	0,0000 0000 0000 0000	0,0281 3696 6805 9756	31
32	6132 2968 2719 7684	0556 2624 4178 4226	0278 1599 9901 7541	32
33	6408 4489 0660 0037	0000 0000 0000 0000	0273 9857 7478 6361	33
34	6679 9695 1819 2544	0537 8914 2894 2666	0268 9154 7012 5583	34
35	6946 0092 6373 6024	0000 0000 0000 0000	0263 0281 7374 9038	35
36	0,7205 7412 5875 0134	0,0512 5081 8908 8729	0,0256 2865 6024 5634	36
37	7458 2803 0706 0855	0000 0000 0000 0000	0248 6444 9476 6087	37
38	7702 7578 2289 7284	0480 4436 3685 0143	0240 1861 1329 4415	38
39	7938 4170 5936 0541	0000 0000 0000 0000	0231 0151 6420 7010	39
40	8164 5398 5973 2476	0442 1157 9271 8785	0221 0982 0654 4328	40
41	0,8380 3441 3442 2522	0,0000 0000 0000 0000	0,0210 3802 4941 8486	41
42	8585 0673 6869 7118	0398 0243 3886 5289	0198 9651 7645 3719	42
43	8778 0917 5213 6144	0000 0000 0000 0000	0186 9939 7366 2517	43
44	8958 8581 9535 2541	0348 7441 1883 1228	0174 4289 1382 3230	44
45	9126 7195 3623 1810	0000 0000 0000 0000	0161 1881 6395 8325	45
46	0,9281 0395 4009 1472	0,0294 9176 8429 9168	0,0147 3865 4122 6756	46
47	9421 3832 9855 0946	0000 0000 0000 0000	0133 2504 6356 8305	47
48	9547 4116 0338 7456	0237 2470 6260 3075	0118 7214 6910 7190	48
49	9658 6296 1227 4941	0000 0000 0000 0000	0103 6169 1593 0595	49
50	9754 5027 8907 3525	0176 4852 6878 7099	0088 0962 6738 8255	50
51	0,9834 8246 2813 1216	0,0000 0000 0000 0000	0,0072 5644 1387 3858	51
52	9899 6173 7980 7506	0113 4311 5798 0903	0056 9675 3172 2434	52
53	9948 5480 1150 8489	0000 0000 0000 0000	0040 7539 2449 5906	53
54	9980 8963 1444 4943	0048 9949 8025 6472	0023 9176 8277 6597	54
55	9996 8277 3261 4886	0000 0000 0000 0000	0008 5443 0248 1755	55
Remainder terms		$G_{54} = 0{,}475 \times 10^{-14}$	$K_{84} = 0{,}756 \times 10^{-26}$	

No.	Nodes			Weights of Gaussian quadratures			Weights of improved quadratures			No.
	Mantissa		Order	Mantissa		Order	Mantissa		Order	
35	4164 7176 1543 7101		0	0000 0000 0000 0000		0	7230 5873 5142 5755		− 5	35
36	4351 8247 2576 5257		0	7206 3111 2664 1272		− 4	7206 0154 3654 7453		− 5	36
37	4535 1241 3106 2211		0	0000 0000 0000 0000		0	7147 7577 0173 0217		− 5	37
40	4717 6236 6603 3705		0	7075 4130 1613 5354		− 4	7075 7141 1704 3073		− 5	40
41	5100 7151 1163 0603		0	0000 0000 0000 0000		0	7007 1373 2164 4406		− 5	41
42	5260 0731 1136 2061		0	6705 1001 0634 2635		− 4	6704 5651 2752 3053		− 5	42
43	5435 0585 6150 5701		0	0000 0000 0000 0000		0	6567 4401 4325 7465		− 5	43
44	5607 3613 6017 5032		0	6436 6140 7433 2620		− 4	6437 1457 7356 4354		− 5	44
45	5756 7225 7605 1235		0	0000 0000 0000 0000		0	6273 0207 5253 4366		− 5	45
46	6123 0313 1335 2717		0	6114 5052 5310 6132		− 4	6114 1255 6117 4032		− 5	46
47	6263 4465 6123 1072		0	0000 0000 0000 0000		0	5723 7544 1407 5064		− 5	47
50	6420 1440 7007 0646		0	5521 3463 3423 6125		− 4	5521 7647 4503 1350		− 5	50
51	6550 4554 2752 4062		0	0000 0000 0000 0000		0	5305 3757 5062 0040		− 5	51
52	6674 3430 7544 0107		0	5060 3740 3730 0142		− 4	5057 7005 4456 5307		− 5	52
53	7013 4032 1072 5724		0	0000 0000 0000 0000		0	4622 7572 4111 1440		− 5	53
54	7125 4305 7175 5244		0	4355 4170 5201 2575		− 4	4356 2145 0021 6273		− 5	54
55	7232 2336 4014 4671		0	0000 0000 0000 0000		0	4100 5217 1622 6176		− 5	55
56	7331 4070 4034 0760		0	7431 4270 4720 5734		− 5	7427 5145 2706 4641		− 6	56
57	7422 7772 2060 4443		0	0000 0000 0000 0000		0	6645 0513 4507 7521		− 6	57
60	7506 4752 5274 6315		0	6045 5120 5375 4346		− 5	6050 1544 5334 0007		− 6	60
61	7564 1313 4156 1712		0	0000 0000 0000 0000		0	5234 2025 5027 7733		− 6	61
62	7633 3434 0154 3771		0	4411 1644 5610 4107		− 5	4405 3100 5356 6437		− 6	62
63	7674 2601 5537 4316		0	0000 0000 0000 0000		0	7334 3561 1615 5123		− 7	63
64	7726 7041 7216 5454		0	5635 4171 6614 6557		− 6	5652 5724 2213 4712		− 7	64
65	7752 7315 7063 3043		0	0000 0000 0000 0000		0	4130 5836 5120 4646		− 7	65
66	7770 1315 2530 3742		0	5010 5767 7113 5570		− 7	4713 7466 4203 5614		−10	66
67	7776 5465 6430 4135		0	0000 0000 0000 0000		0	6777 5727 7077 4113		−12	67

No.	Nodes	Weights of Gaussian quadratures	Weights of improved quadratures	No.
I	0,0002 9523 6267 7709	0,0000 0000 0000 0000	0,0007 9536 8020 3534	I
2	00I7 7875 I2I3 0228	0045 62I4 I296 5473	0022 2746 8I48 3462	2
3	0047 9I37 8I96 48I5	0000 0000 0000 0000	0037 9540 0526 5520	3
4	0093 484I 73I4 5636	0I05 6605 6296 3856	0053 0592 942I 0857	4
5	0I53 8432 6320 2028	0000 0000 0000 0000	0067 6I33 2665 68I6	5
6	0,0228 7035 9685 5309	0,0I64 507I 389I I522	0,0082 I244 2899 920I	6
7	03I8 0952 4837 5749	0000 0000 0000 0000	0096 6289 0I88 4I29	7
8	042I 8348 6803 9840	022I 3646 7879 502I	0II0 765I I690 9206	8
9	0539 4565 556I 594I	0000 0000 0000 0000	0I24 4063 7I80 48I3	9
I0	0670 5373 87I2 8025	0275 5367 2837 8584	0I37 7I07 04I9 39I7	I0
II	0,08I4 7789 5790 3303	0,0000 0000 0000 0000	0,0I50 7090 6957 6473	II
I2	097I 793I 454I 4I04	0326 3646 I983 4998	0I63 2243 9445 5I02	I2
I3	II4I 0327 6378 288I	0000 0000 0000 0000	0I75 I6I6 II75 9220	I3
I4	I32I 9456 0993 I84I	0373 23I0 7II7 2844	0I86 5838 7533 0090	I4
I5	I5I4 034I 4027 55II	0000 0000 0000 0000	0I97 50I8 I0I9 I555	I5
I6	0,I7I6 7445 2980 5675	0,04I5 5670 86I4 4506	0,0207 8077 02II 4652	I6
I7	I929 42I9 II35 0579	0000 0000 0000 0000	02I7 4548 273I 7674	I7
I8	2I5I 3976 4094 299I	0452 8587 2I96 5I64	0226 4I08 7485 8892	I8
I9	2382 0276 7278 5835	0000 0000 0000 0000	0234 7358 737I 7980	I9
20	2620 6288 7522 4409	0484 6532 8998 9650	0242 3406 I9I5 9473	20
2I	0,2866 4535 4034 4464	0,0000 0000 0000 0000	0,0249 I8I7 2453 8896	2I
22	3II8 7424 I955 4606	05I0 5648 3789 0304	0255 272I 8482 8443	22
23	3376 7468 I47I 0898	0000 0000 0000 0000	0260 6076 7I20 I435	23
24	3639 69I8 6I82 4II0	0530 2788 296I 4232	0265 I464 I239 2794	24
25	3906 766I 7III 2298	0000 0000 0000 0000	0268 8660 4I04 537I	25
26	0,4I77 I535 8933 3096	0,0543 5559 6I29 I47I	0,027I 7738 9938 2I9I	26
27	4450 04I7 5354 6336	0000 0000 0000 0000	0273 8644 8422 3474	27
28	4724 6035 5057 9829	0550 2350 6508 2376	0275 II88 7404 5386	28
29	0,5000 0000 0000 0000	0,0000 0000 0000 0000	0,0275 5350 7649 8889	29

No.	Nodes Mantissa	Order	Weights of Gaussian quadratures Mantissa	Order	Weights of improved quadratures Mantissa	Order	No.
I	4654 4760 7255 3650	-I3	0000 0000 0000 0000	0	64I0 0076 4257 2600	-I2	I
2	7222 2374 4464 3505	-II	4527 7003 6530 7056	- 7	4437 5266 668I 5572	-I0	2
3	4720 0376 2500 65I3	- 7	0000 0000 0000 0000	0	76I3 6I0I I060 I327	-I0	3
4	4622 5032 52I2 5222	- 6	532I 6500 4366 3450	- 6	5335 6534 5265 I074	- 7	4
5	7700 72I2 4652 20I6	- 6	0000 0000 0000 0000	0	6730 7053 255I 3333	- 7	5
6	5665 5236 67I3 2444	- 5	4I54 I645 6I2I 0467	- 5	4I50 6573 3263 5046	- 6	6
7	4044 5264 2340 7426	- 4	0000 0000 0000 0000	0	4745 043I 2540 25I2	- 6	7
I0	53I4 4227 4I0I 037I	- 4	5525 36II 3343 226I	- 5	5527 5I0I 6735 35I5	- 6	I0
II	67I7 3036 5I45 76I4	- 4	0000 0000 0000 0000	0	6275 I720 3523 4I55	- 6	II
I2	4225 I570 3554 2243	- 3	7033 4075 3562 2447	- 5	7032 00I6 2I24 I305	- 6	I2
I3	5I55 6742 0342 0603	- 3	0000 0000 0000 0000	0	7556 5767 I004 20I7	- 6	I3
I4	6I60 2762 6346 6250	- 3	4I32 67I7 4340 2475	- 4	4I33 3242 7470 0I02	- 5	I4
I5	7232 7372 4064 5026	- 3	0000 0000 0000 0000	0	4367 70I5 33I5 6627	- 5	I5
I6	4I65 7002 64I3 I377	- 2	46I6 0035 2I64 I357	- 4	46I5 457I 4200 I403	- 5	I6
I7	4660 4577 0760 0432	- 2	0000 0000 0000 0000	0	5034 544I 5442 370I	- 5	I7
20	5374 5555 4I07 0337	- 2	5243 3536 0I55 5400	- 4	5243 6I57 0343 2265	- 5	20
2I	6I3I I243 2074 3II5	- 2	0000 0000 0000 0000	0	544I 7543 3I67 6630	- 5	2I
22	6704 668I I266 5253	- 2	5627 6655 60I6 0303	- 4	5627 47I5 2230 3522	- 5	22
23	7476 5555 072I 6344	- 2	0000 0000 0000 0000	0	6004 5656 205I 64I4	- 5	23
24	4I42 6433 2554 7364	- I	6I50 I624 54I2 2637	- 4	6I50 3202 730I 5I73	- 5	24
25	4454 I456 0244 0657	- I	0000 0000 0000 0000	0	6302 046I 7607 7525	- 5	25
26	4772 6773 0265 I665	- I	6422 0232 4037 3I25	- 4	642I 7I65 0425 25I5	- 5	26
27	53I6 I662 I063 30I2	- I	0000 0000 0000 0000	0	6527 6543 6372 5607	- 5	27
30	5645 5053 2354 7565	- I	6623 I703 7430 3337	- 4	6623 2473 477I 626I	- 5	30
3I	6200 3303 767I 5772	- I	0000 0000 0000 0000	0	6704 05I3 5250 5447	- 5	3I
32	6535 73II 4664 763I	- I	6752 I77I I672 5427	- 4	6752 I436 0763 2624	- 5	32
33	7075 3626 2567 3224	- I	0000 0000 0000 0000	- 0	7005 46I3 4247 557I	- 5	33
34	7436 3I22 6664 I333	- I	7026 0I24 0366 64I4	- 4	7026 0234 05I6 5526	- 5	34
35	4000 0000 0000 0000	0	0000 0000 0000 0000	0	7033 3744 6I76 I053	- 5	35

No.	Nodes	Weights of Gaussian quadratures	Weights of improved quadratures	No.
30	0,5275 3964 4942 0I7I	0,0550 2350 6508 2376	0,0275 II88 7404 5386	30
3I	5549 9582 4645 3664	0000 0000 0000 0000	0273 8644 8422 3474	3I
32	5822 8464 I066 6904	0548 5559 6I29 I47I	027I 7738 9938 2I9I	32
33	0,6098 2338 2888 7702	0,0000 0000 0000 0000	0,0268 8660 4I04 537I	33
34	6360 308I 38I7 5890	0530 2788 296I 4232	0265 I464 I239 2794	34
35	6623 253I 8528 9I02	0000 0000 0000 0000	0260 6076 7I20 I435	35
36	688I 2575 8044 5394	05I0 5648 3789 0304	0255 272I 8482 8443	36
37	7I33 5464 5965 5536	0000 0000 0000 0000	0249 I8I7 2453 8896	37
38	0,7379 37II 2477 559I	0,0484 6532 8998 9650	0,0242 3406 I9I5 9473	38
39	76I7 9723 272I 4I65	0000 0000 0000 0000	0234 7358 737I 7980	39
40	7848 6023 5905 7009	0452 8587 2I96 5I64	0226 4I08 7485 8892	40
4I	8070 5780 8864 942I	0000 0000 0000 0000	02I7 4348 273I 7674	4I
42	8283 2554 70I9 4325	04I5 5670 86I4 4506	0207 8077 02II 4652	42
43	0,8485 9658 5972 4489	0,0000 0000 0000 0000	0,0I97 50I8 I0I9 I555	43
44	8678 0543 9006 8I59	0373 23I0 7II7 2844	0I86 5838 7533 0090	44
45	8858 9672 362I 7I69	0000 0000 0000 0000	0I75 I6I6 II75 9220	45
46	9028 2068 5458 5896	0326 3646 I983 4998	0I63 2243 9445 5I02	46
47	9I85 22I0 4209 6697	0000 0000 0000 0000	0I50 7090 6957 6473	47
48	0,9329 4626 I287 I975	0,0275 5367 2837 8584	0,0I37 7I07 04I9 39I7	48
49	9460 5434 4438 4059	0000 0000 0000 0000	0I24 4063 7I80 48I3	49
50	9578 I65I 3I96 0660	022I 3646 7379 502I	0II0 765I I690 9206	50
5I	968I 9047 5I62 425I	0000 0000 0000 0000	0096 6289 0I88 4I29	5I
52	977I 2964 03I4 469I	0I64 507I 389I I522	0082 I244 2899 920I	52
53	0,9846 I567 3679 7972	0,0000 0000 0000 0000	0,0067 6I33 2665 68I6	53
54	9906 5I58 2685 4364	0I05 6605 6296 3856	0053 0592 942I 0857	54
55	9952 0862 I803 5I85	0000 0000 0000 0000	0037 9540 0526 5520	55
56	9982 2I24 8786 9772	0045 62I4 I296 5473	0022 2746 8I48 3462	56
57	9997 0476 3732 229I	0000 0000 0000 0000	0007 9536 8020 3534	57
Remainder terms	$G_{56} = 0,123 \times 10^{-14}$	$K_{86} = 0,880 \times 10^{-27}$		

No.	Nodes Mantissa	Order	Weights of Gaussian quadratures Mantissa	Order	Weights of improved quadratures Mantissa	Order	No.
36	4I60 6326 4445 7222	0	7026 0I24 0366 64I4	− 4	7026 0234 05I6 5526	− 5	36
37	434I 2064 6504 2265	0	0000 0000 0000 0000	0	7005 46I3 4247 557I	− 5	37
40	452I 0233 I445 4063	0	6752 I77I I672 5427	− 4	6752 I436 0763 2624	− 5	40
4I	4677 6236 0043 I002	0	0000 0000 0000 0000	0	6704 05I3 5250 5447	− 5	4I
42	5055 I352 26II 4I05	0	6623 I703 7430 3337	− 4	6623 2473 477I 626I	− 5	42
43	5230 7046 7346 2372	0	0000 0000 0000 0000	0	6527 6543 6372 5607	− 5	43
44	5402 4402 3645 3045	0	6422 0232 4037 3125	− 4	642I 7I65 0425 25I5	− 5	44
45	555I 7I50 7655 7450	0	0000 0000 0000 0000	0	6302 046I 7607 7525	− 5	45
46	57I6 4562 25II 4205	0	6I50 I624 54I2 2637	− 4	6I50 3202 730I 5I73	− 5	46
47	6060 2444 56I3 4306	0	0000 0000 0000 0000	0	6004 5656 205I 64I4	− 5	47
50	62I6 623I 5522 2525	0	5627 6655 60I6 0303	− 4	5627 47I5 2230 3522	− 5	50
5I	635I 5527 I360 7I54	0	0000 0000 0000 0000	0	544I 7543 3I67 6630	− 5	5I
52	6500 6444 4756 I7I0	0	5243 3536 0I55 5400	− 4	5243 6I57 0343 2265	− 5	52
53	6623 6640 I603 767I	0	0000 0000 0000 0000	0	5034 544I 5442 370I	− 5	53
54	6742 4I77 2275 I500	0	46I6 0035 2I64 I357	− 4	46I5 457I 4200 I408	− 5	54
55	7054 5040 537I 3275	0	0000 0000 0000 0000	0	4367 70I5 33I5 6627	− 5	55
56	7I6I 750I 5I43 II52	0	4I32 67I7 4340 2475	− 4	4I33 3242 7470 0I02	− 5	56
57	7262 2I03 5743 57I7	0	0000 0000 0000 0000	0	7556 5767 I004 20I7	− 6	57
60	7355 2620 7422 3553	0	7033 4075 3562 2447	− 5	7032 00I6 2I24 I305	− 6	60
6I	7443 0236 053I 5007	0	0000 0000 0000 0000	0	6275 I720 3523 4I55	− 6	6I
62	7523 I566 4I73 7360	0	5525 36II 3843 226I	− 5	5527 5I0I 6735 35I3	− 6	62
63	7575 5524 566I 74I6	0	0000 0000 0000 0000	0	4745 043I 2540 25I2	− 6	63
64	7642 2453 022I 5I26	0	4I54 I645 6I2I 0467	− 5	4I50 6573 3263 5046	− 6	64
65	7700 7705 653I 2557	0	0000 0000 0000 0000	0	6730 7053 255I 3333	− 7	65
66	773I 5527 4525 6525	0	532I 6500 4866 3450	− 6	5335 6534 5265 I074	− 7	66
67	7754 2776 0065 3745	0	0000 0000 0000 0000	0	76I3 6I0I I060 I827	−I0	67
70	7770 5555 4033 3I34	0	4527 7003 6580 7056	− 7	4437 5266 663I 5572	−I0	70
7I	7776 6246 6036 I245	0	0000 0000 0000 0000	0	64I0 0076 4257 2600	−I2	7I

No.	Nodes	Weights of Gaussian quadratures	Weights of improved quadratures	No.
I	0,0002 7566 6058 9746	0,0000 0000 0000 0000	0,0007 425I 6I52 06I8	I
2	00I6 6027 8869 70I7	0042 5845 I939 3732	0020 7887 I223 6967	2
3	0044 7229 7I38 5972	0000 0000 0000 0000	0035 43I8 2293 95I0	3
4	0087 2724 7369 2934	0098 6604 2528 06I4	0049 5487 6778 9898	4
5	0I43 6438 I449 3465	0000 0000 0000 0000	0063 I5I4 8I47 3I03	5
6	0,02I3 5720 2II0 956I	0,0I53 7024 6I0I 0468	0,0076 7245 2086 8833	6
7	0297 I054 I998 5I36	0000 0000 0000 0000	0090 3I9I 3073 27I6	7
8	0394 0988 3523 4706	0207 0I03 I259 34I4	0I03 5897 0I78 9675	8
9	0504 I286 4703 6I02	0000 0000 0000 0000	0II6 4040 7508 4458	9
I0	0626 8I09 7589 9486	0257 974I 345I 2490	0I28 9248 8689 07I0	I0
II	0,076I 9004 0II9 5II6	0,0000 0000 0000 0000	0,0I4I 20I7 5782 343I	II
I2	0909 0725 6I92 3738	0306 0I54 5328 5896	0I53 0563 952I 0467	I2
I3	I067 8340 0209 7523	0000 0000 0000 0000	0I64 88I5 2469 5952	I3
I4	I237 6857 4I32 76I4	0350 5896 6627 5256	0I75 2548 3440 0I55	I4
I5	I4I8 2066 7336 2087	0000 0000 0000 0000	0I85 7I04 0I0I I026	I5
I6	0,I608 9273 II98 6567	0,039I I9I6 3567 88I9	0,0I95 6298 9866 2498	I6
I7	I809 257I 5760 5764	0000 0000 0000 0000	0204 9273 6942 7597	I7
I8	20I8 59I0 I480 886I	0427 36I2 8683 0863	02I3 6506 9337 4264	I8
I9	2236 3782 II49 3402	0000 0000 0000 0000	022I 8278 5375 0482	I9
20	2462 0352 2437 8862	0458 6887 8569 6294	0229 37I3 7878 4558	20
2I	0,2694 8847 7305 883I	0,0000 0000 0000 0000	0,0236 2III 0468 80II	2I
22	2934 2355 59I2 9957	0484 8I9I 7047 2043	0242 3847 0860 I70I	22
23	3I79 448I 0643 3III	0000 0000 0000 0000	0247 9207 3576 4099	23
24	3429 84I8 I066 I800	0505 4563 6879 9575	0252 75I5 7264 4I56	24
25	3684 6905 I852 39I4	0000 0000 0000 0000	0256 8I94 II33 204I	25
26	0,3943 2385 69I6 9995	0,0520 3665 5038 8647	0,0260 I608 6592 0009	26
27	4204 7826 0997 7390	0000 0000 0000 0000	0262 8094 9206 3757	27
28	4468 6088 4933 6604	0529 3807 7548 6605	0264 7I22 3I67 9465	28
29	4733 94II 3897 6753	0000 0000 0000 0000	0265 82II I408 28I4	29
30	0,5000 0000 0000 0000	0,0532 3969 0859 I57I	0,0266 I767 895I 5058	30

No.	Nodes		Weights of Gaussian quadratures		Weights of improved quadratures		No.
	Mantissa	Order	Mantissa	Order	Mantissa	Order	
I	44I0 3505 5234 7I6I	-I3	0000 0000 0000 0000	0	6052 2552 2700 I6I5	-I2	I
2	663I 6666 7I22 6366	-II	4270 5I73 7I75 2502	- 7	4203 6653 726I 5I70	-I0	2
3	4450 6I30 7626 0342	- 7	0000 0000 0000 0000	0	7203 2274 0470 3565	-I0	3
4	4357 6272 4044 0377	- 6	5032 2456 3776 7432	- 6	5045 6204 6705 I04I	- 7	4
5	7265 4225 I065 356I	- 6	0000 0000 0000 0000	0	6356 75I5 3030 6I02	- 7	5
6	5357 25I4 4463 4I04	- 5	7675 I574 0527 4II4	- 6	7666 446I 3I30 52I5	- 7	6
7	7466 I605 6I65 6222	- 5	0000 0000 0000 0000	0	4477 5226 6424 376I	- 6	7
I0	5026 6I02 0337 2650	- 4	523I 2465 420I I465	- 5	5233 4253 42I6 I720	- 6	I0
II	6347 6670 2463 0742	- 4	0000 0000 0000 0000	0	5753 3550 3076 2624	- 6	II
I2	4005 7362 400I 3573	- 3	6465 2430 7060 3056	- 5	6463 5404 2442 325I	- 6	I2
I3	4700 4606 05I4 0373	- 3	0000 0000 0000 0000	0	7I65 4II7 28I3 553I	- 6	I3
I4	5642 6625 3050 2720	- 3	7653 0027 43II 2230	- 5	7654 220I 2277 6023	- 6	I4
I5	6653 050I 2724 0245	- 3	0000 0000 0000 0000	0	4I52 45I5 7206 7460	- 5	I5
I6	7727 5I40 6426 0720	- 3	437I 4775 5430 6I47	- 4	437I 0682 2I6I 3053	- 5	I6
I7	4423 4557 7006 6I64	- 2	0000 0000 0000 0000	0	4602 III3 I706 0720	- 5	I7
20	5II4 0420 3252 3076	- 2	5003 5552 4053 6036	- 4	5004 I220 I540 0056	- 5	20
2I	5624 2227 20I4 47I4	- 2	0000 0000 0000 0000	0	5I76 0I42 2763 0234	- 5	2I
22	6353 2046 7702 7262	- 2	5360 6024 I4I5 I364	- 4	5360 27I4 205I 3664	- 5	22
23	7I20 0520 0777 030I	- 2	0000 0000 0000 0000	0	5533 4254 I560 5I45	- 5	23
24	770I 6306 550I 7565	- 2	5676 040I 2534 2645	- 4	5676 3252 I082 4I56	- 5	24
25	4237 5I44 6225 5357	- I	0000 0000 0000 0000	0	6030 04I7 0752 6663	- 5	25
26	4543 5634 6020 5560	- I	6I5I 287I 4052 2I34	- 4	6I50 770I 75I4 42II	- 5	26
27	5054 4630 6005 I427	- I	0000 0000 0000 0000	0	626I 4277 II53 7032	- 5	27
30	537I 5637 3027 I374	- I	6360 436I 0532 6664	- 4	6360 6730 5664 I36I	- 5	30
3I	57I2 377I 575I 3232	- I	0000 0000 0000 0000	0	6446 I357 I207 5042	- 5	3I
32	6236 232I 0257 5560	- I	6522 2I43 I625 63I6	- 4	652I 7660 I052 2073	- 5	32
33	6564 4355 I565 2027	- I	0000 0000 0000 0000	0	6564 5445 I262 632I	- 5	33
34	7II4 5363 I324 6652	- I	66I5 2630 77I0 670I	- 4	66I5 5055 5625 I6II	- 5	34
35	7446 0266 465I 7I53	- I	0000 0000 0000 0000	0	6634 I272 3I20 3702	- 5	35
36	4000 0000 0000 0000	0	664I 0734 5426 6574	- 4	6640 652I 4457 I332	- 5	36

No.	Nodes	Weights of Gaussian quadratures	Weights of improved quadratures	No.
3I	0,5266 0588 6I02 3247	0,0000 0000 0000 0000	0,0265 82II I408 28I4	3I
32	553I 39II 5066 3396	0529 3807 7548 6605	0264 7I22 3I67 9465	32
33	5795 2I73 9002 26I0	0000 0000 0000 0000	0262 8094 9206 3757	33
34	6056 76I4 3083 0005	0520 3665 5038 8647	0260 I608 6592 0009	34
35	0,63I5 3094 8I47 6086	0,0000 0000 0000 0000	0,0256 8I94 II33 204I	35
36	6570 I58I 8933 8200	0505 4563 6879 9575	0252 75I5 7264 4I56	36
37	6820 5568 9356 6889	0000 0000 0000 0000	0247 9207 3576 4099	37
38	7065 7644 4087 0043	0484 8I9I 7047 2043	0242 3847 0860 I70I	38
39	7305 II52 2694 II69	0000 0000 0000 0000	0236 2III 0468 80II	39
40	0,7537 9647 7562 II38	0,0458 6887 8569 6294	0,0229 37I3 7878 4558	40
4I	7763 62I7 8850 6598	0000 0000 0000 0000	022I 8278 5375 0482	4I
42	798I 4089 8569 II39	0427 36I2 8683 0863	02I3 6506 9387 4264	42
43	8I90 7428 4239 4236	0000 0000 0000 0000	0204 9273 6942 7597	43
44	839I 0726 880I 3433	039I I9I6 3567 88I9	0I95 6298 9866 2498	44
45	0,858I 7933 2663 79I3	0,0000 0000 0000 0000	0,0I85 7I04 0I0I I026	45
46	8762 3I42 5867 2386	0350 5896 6627 5256	0I75 2548 3440 0I55	46
47	8932 I659 9790 2477	0000 0000 0000 0000	0I64 38I5 2469 5952	47
48	9090 9274 3807 6262	0306 0I54 5328 5896	0I58 0563 952I 0467	48
49	9238 0995 9880 4884	0000 0000 0000 0000	0I4I 20I7 5782 343I	49
50	0,9373 I890 2460 05I4	0,0257 974I 345I 2490	0,0I28 9248 8689 07I0	50
5I	9495 87I3 5296 3898	0000 0000 0000 0000	0II6 4040 7508 4458	5I
52	9605 90II 6476 5294	0207 0I03 I259 34I4	0I03 5897 0I78 9675	52
53	9702 8945 800I 4864	0000 0000 0000 0000	0090 3I9I 3073 27I6	53
54	9786 4279 7889 0439	0I53 7024 6I0I 0468	0076 7245 2086 8833	54
55	0,9856 356I 8550 6535	0,0000 0000 0000 0000	0,0063 I5I4 8I47 3I03	55
56	99I2 7275 2630 7066	0098 6604 2528 06I4	0049 5487 6778 9398	56
57	9955 2770 286I 4028	0000 0000 0000 0000	0035 48I8 2293 95I0	57
58	9983 3972 II30 2983	0042 5845 I939 3732	0020 7887 I223 6967	58
59	9997 2433 394I 0254	0000 0000 0000 0000	0007 425I 6I52 06I8	59
Remainder terms		$G_{58} = 0{,}3I9 \times 10^{-I5}$	$K_{90} = 0{,}I06 \times 10^{-27}$	

No.	Nodes Mantissa	Order	Weights of Gaussian quadratures Mantissa	Order	Weights of improved quadratures Mantissa	Order	No.
37	4I54 7644 5453 03I2	0	0000 0000 0000 0000	0	6634 I272 3I20 3702	- 5	37
40	433I 5206 3225 4452	0	66I5 2630 77I0 670I	- 4	66I5 5055 5625 I6II	- 5	40
4I	4505 56II 3I05 2764	0	0000 0000 0000 0000	0	6564 5445 I262 632I	- 5	4I
42	4660 6627 3650 II07	0	6522 2I43 I625 63I6	- 4	652I 7660 I052 2073	- 5	42
43	5032 6003 I0I3 2262	0	0000 0000 0000 0000	0	6446 I857 I207 5042	- 5	43
44	5203 I060 2364 320I	0	6360 436I 0532 6664	- 4	6360 6730 5664 I36I	- 5	44
45	535I 5463 4775 3I64	0	0000 0000 0000 0000	0	626I 4277 II53 7032	- 5	45
46	55I6 I06I 4767 5I07	0	6I5I 237I 4052 2I34	- 4	6I50 770I 75I4 42II	- 5	46
47	5660 I8I5 4665 I2I0	0	0000 0000 0000 0000	0	6030 04I7 0752 6663	- 5	47
50	60I7 43I6 2457 4042	0	5676 040I 2534 2645	- 4	5676 3252 I032 4I56	- 5	50
5I	6I53 7653 7600 I7I7	0	0000 0000 0000 0000	0	5533 4254 I560 5I45	- 5	5I
52	6305 I366 20I7 2I23	0	5360 6024 I4I5 I364	- 4	5360 27I4 205I 3664	- 5	52
53	6482 7332 I374 66I4	0	0000 0000 0000 0000	0	5I76 0I42 2763 0234	- 5	53
54	6554 7673 7I25 3I60	0	5003 5552 4053 6036	- 4	5004 I220 I540 0056	- 5	54
55	6673 0644 0I76 2342	0	0000 0000 0000 0000	0	4602 III3 I706 0720	- 5	55
56	7005 0263 7I35 I705	0	437I 4775 5430 6I47	- 4	437I 0632 2I6I 3053	- 5	56
57	7II2 4727 6505 3753	0	0000 0000 0000 0000	0	4I52 45I5 7206 7460	- 5	57
60	72I3 5II5 2472 7505	0	7653 0027 43II 2230	- 5	7654 220I 2277 6023	- 6	60
6I	7307 73I7 I726 3740	0	0000 0000 0000 0000	0	7I65 4II7 23I3 553I	- 6	6I
62	7377 204I 5377 6420	0	6465 2430 7060 3056	- 5	6463 5404 2442 325I	- 6	62
63	746I 4044 3654 634I	0	0000 0000 0000 0000	0	5753 3550 3076 2624	- 6	63
64	7536 4473 6762 0245	0	523I 2465 420I I465	- 5	5233 4253 42I6 I720	- 6	64
65	7606 2343 6434 2433	0	0000 0000 0000 0000	0	4477 5226 6424 376I	- 6	65
66	7650 4I25 4666 3075	0	7675 I574 0527 4II4	- 6	7666 446I 3I30 52I5	- 7	66
67	7705 I235 5267 I242	0	0000 0000 0000 0000	0	6356 75I5 3030 6I02	- 7	67
70	7734 20I5 0537 3374	0	5032 2456 3776 7432	- 6	5045 6204 6705 I04I	- 7	70
7I	7755 5347 2340 6476	0	0000 0000 0000 0000	0	7203 2274 0470 3565	-I0	7I
72	777I I46I III0 655I	0	4270 5I73 7I75 2502	- 7	4203 6653 726I 5I70	-I0	72
73	7776 6757 0564 5306	0	0000 0000 0000 0000	0	6052 2552 2700 I6I5	-I2	73

No.	Nodes	Weights of Gaussian quadratures	Weights of improved quadratures	No.
I	0,0002 5779 4974·7547	0,0000 0000 0000 0000	0,0006 9450 6849 8385	I
2	00I5 5325 7962 6752	0039 8409 6248 0833	00I9 4523 0563 5499	2
3	004I 8450 I564 7977	0000 0000 0000 0000	0033 I535 I957 9656	3
4	008I 6593 8360 I264	0092 3323 4I55 5455	0046 8663 9829 7589	4
5	0I34 4I83 8749 4369	0000 0000 0000 0000	0059 II50 7626 7482	5
6	0,0199 8906 75I5 8462	0,0I43 9235 394I 66I7	0,007I 8486 4753 5229	6
7	0278 I277 7625 7200	0000 0000 0000 0000	0084 6044 4594 5266	7
8	0368 9997 6285 3628	0I93 9959 6284 8I35	·0097 0707 0596 97I2	8
9	0472 I334 6I50 046I	0000 0000 0000 0000	0I09 I40I 79I0 8046	9
IO	0587 I973 2I03 9737	0242 0I33 64I5 2970	0I20 9558 I039 0403	IO
II	0,07I3 9738 3226 9694	0,0000 0000 0000 0000	0,0I32 5497 744I I666	II
I2	0852 I7II 8808 6I58	0287 4657 8I08 8095	0I43 7702 4382 5206	I2
I3	I00I 3608 2089 0805	0000 0000 0000 0000	0I54 5362 878I I939	I3
I4	II6I II28 3947 5869	0329 87II 494I 0902	0I64 9072 3528 74I9	I4
I5	I33I 0496 8773 3866	0000 0000 0000 0000	0I74 8966 90I4 0300	I5
I6	0,I5I0 7475 2603 342I	0,0368 7798 7368 8526	0,0I84 4II8 2325 9I06	I6
I7	I699 6946 7936 6865	0000 0000 0000 0000	0I93 3947 28I2 3638	I7
I8	I897 3690 8505 3786	0403 7794 76I4 7I0I	020I 8726 9475 7680	I8
I9	2I03 2738 2086 8I92	0000 0000 0000 0000	0209 8490 5I07 582I	I9
20	23I6 8792 5928 9900	0434 4989 3600 54I5	02I7 2626 9850 6780	20
2I	0,2537 5976 6069 II07	0,0000 0000 0000 0000	0,0224 0740 0066 58I3	2I
22	2764 83II 5230 9554	0460 6I26 III8 893I	0230 296I 9I35 5035	22
23	2997 9937 2584 8028	0000 0000 0000 0000	0235 9277 3284 6496	23
24	3236 4763 7234 5609	048I 8436 8587 322I	0240 9293 0878 5436	24
25	3479 6339 8863 I875	0000 0000 0000 0000	0245 277I 7277 5I49	25
26	0,3726 8I53 69I6 055I	0,0497 967I 0293 3976	0,0248 9784 I7I3 537I	26
27	3977 3744 I658 845I	0000 0000 0000 0000	0252 0296 070I 39I2	27
28	4230 6504 3I95 7082	0508 8II9 4874 2028	0254 4089 7949 3748	28
29	4485 9653 I0I6 63I5	0000 0000 0000 0000	0256 I077 3924 6294	29
30	4742 6407 8722 34I2	05I4 2632 6446 7797	0257 I306 4268 7295	30
3I	0,5000 0000 0000 0000	0,0000 0000 0000 0000	0,0257 4736 47I4 7258	3I

n = 36

No.	Nodes Mantissa	Order	Weights of Gaussian quadratures Mantissa	Order	Weights of improved quadratures Mantissa	Order	No.
I	4162 4247 7051 4625	−13	0000 0000 0000 0000	0	5540 7620 6432 5711	−12	I
2	6271 3260 2334 5515	−11	4050 6405 4267 4441	− 7	7757 3433 2352 4047	−11	2
3	4221 7044 5367 4221	− 7	0000 0000 0000 0000	0	6624 3140 1677 2002	−10	3
4	4134 5155 4312 5117	− 6	4564 3375 5360 0015	− 6	4576 7364 1565 0212	− 7	4
5	6703 5450 3516 6211	− 6	0000 0000 0000 0000	0	6033 2521 7552 2747	− 7	5
6	5074 0034 7204 1122	− 5	7274 6750 2767 5457	− 6	7266 7403 4443 7611	− 7	6
7	7075 3637 2240 3506	− 5	0000 0000 0000 0000	0	4251 6655 1587 2613	− 6	7
I0	4562 2155 7536 3244	− 4	4756 5746 7462 5000	− 5	4760 5147 5404 0312	− 6	I0
II	6026 1310 0646 1406	− 4	0000 0000 0000 0000	0	5455 0265 3772 5732	− 6	II
I2	7410 2032 0404 1587	− 4	6144 0741 4330 5317	− 5	6142 6213 1764 4156	− 6	I2
I3	4443 4312 4375 3250	− 3	0000 0000 0000 0000	0	6622 5547 5171 7050	− 6	I3
I4	5350 3120 0453 6443	− 3	7267 6761 4662 6407	− 5	7270 6634 3045 4642	− 6	I4
I5	6321 2045 3334 3402	− 3	0000 0000 0000 0000	0	7723 0467 4341 0153	− 6	I5
I6	7334 5700 5640 1373	− 3	4161 6577 1734 4062	− 4	4161 3615 6174 4517	− 5	I6
I7	4204 6253 1764 2146	− 2	0000 0000 0000 0000	0	4364 3176 4385 6046	− 5	I7
20	4653 1527 0158 5407	− 2	4560 6587 2661 0172	− 4	4561 0766 2772 4325	− 5	20
2I	5340 6171 7206 6446	− 2	0000 0000 0000 0000	0	4746 6720 3200 4017	− 5	2I
22	6044 5144 3067 7675	− 2	5126 1530 6174 0652	− 4	5125 7705 6143 2361	− 5	22
23	6566 0017 5341 0223	− 2	0000 0000 0000 0000	0	5276 4211 1113 1000	− 5	23
24	7323 7631 4057 4353	− 2	5437 4204 0007 3316	− 4	5437 5512 2362 0241	− 5	24
25	4036 6314 6411 6551	− I	0000 0000 0000 0000	0	5570 7671 2362 6752	− 5	25
26	4330 7461 7072 0125	− I	5712 5273 5036 1776	− 4	5712 4234 5015 3644	− 5	26
27	4627 7515 5223 4741	− I	0000 0000 0000 0000	0	6024 2641 5533 7221	− 5	27
30	5133 2444 5053 1502	− I	6125 6370 7502 6342	− 4	6125 7211 6140 2457	− 5	30
3I	5442 4102 1512 2421	− I	0000 0000 0000 0000	0	6216 7131 7432 6702	− 5	3I
32	5755 0035 2234 0504	− I	6277 3642 5050 4526	− 4	6277 3216 7715 0756	− 5	32
33	6272 2075 7412 1616	− I	0000 0000 0000 0000	0	6347 3160 3763 0765	− 5	33
34	6611 5773 1623 3171	− I	6406 4314 5767 2451	− 4	6406 4556 0530 6316	− 5	34
35	7132 7161 6272 0607	− I	0000 0000 0000 0000	0	6484 6657 4334 7436	− 5	35
36	7455 1275 6170 7587	− I	6452 2151 3076 0235	− 4	6452 2064 2072 3407	− 5	36
37	4000 0000 0000 0000	0	0000 0000 0000 0000	0	6456 6043 1320 6706	− 5	37

No.	Nodes	Weights of Gaussian quadratures	Weights of improved quadratures	No.
32	0,5257 3592 I277 6588	0,05I4 2632 6446 7797	0,0257 I306 4268 7295	32
33	55I4 0346 8983 3685	0000 0000 0000 0000	0256 I077 3924 6294	33
34	5769 3495 6804 29I8	0508 8II9 4874 2028	0254 4089 7949 3748	34
35	6022 6255 834I I549	0000 0000 0000 0000	0252 0296 070I 39I2	35
36	6273 I846 3088 9449	0497 967I 0298 3976	0248 9784 I7I3 587I	36
37	0,6520 3660 II36 8I25	0,0000 0000 0000 0000	0,0245 277I 7277 5I49	37
38	6763 5236 2765 439I	048I 8436 8587 322I	0240 9293 0878 5436	38
39	7002 0062 74I5 I972	0000 0000 0000 0000	0235 9277 3284 6496	39
40	7235 I688 4769 0446	0460 6I26 III8 898I	0230 296I 9I35 5035	40
4I	7462 4023 3930 8893	0000 0000 0000 0000	0224 0740 0066 58I3	4I
42	0,7683 I207 407I 0099	0,0434 4989 3600 54I5	0,02I7 2626 9850 6780	42
43	7896 726I 79I3 I808	0000 0000 0000 0000	0209 8490 5I07 582I	43
44	8I02 6309 I494 62I4	0403 7794 76I4 7I0I	020I 8726 9475 7680	44
45	8300 3053 2063 3I35	0000 0000 0000 0000	0I93 3947 28I2 3638	45
46	8489 2524 7396 6579	0368 7798 7368 8526	0I84 4II8 2325 9I06	46
47	0,8668 9503 I226 6I34	0,0000 0000 0000 0000	0,0I74 8966 90I4 0300	47
48	8838 887I 6052 4I3I	0329 87II 494I 0902	0I64 9072 3528 74I9	48
49	8998 639I 79I0 9I95	0000 0000 0000 0000	0I54 5362 878I I989	49
50	9I47 8288 II9I 3842	0287 4657 8I08 8095	0I43 7702 4382 5206	50
5I	9286 026I 6773 0305	0000 0000 0000 0000	0I32 5497 744I I666	5I
52	0,94I2 8026 7896 0263	0,0242 0I33 64I5 2970	0,0I20 9558 I039 0403	52
53	9527 8665 3849 9539	0000 0000 0000 0000	0I09 I40I 79I0 8046	53
54	968I 0002 37I4 6372	0I93 9959 6284 8I35	0097 0707 0596 97I2	54
55	972I 8722 2374 2800	0000 0000 0000 0000	0084 6044 4594 5266	55
56	9800 I098 2484 I538	0I43 9235 394I 66I7	007I 8486 4753 5229	56
57	0,9865 58I6 I250 568I	0,0000 0000 0000 0000	0,0059 II50 7626 7482	57
58	99I8 3406 I639 8736	0092 3823 4I55 5455	0046 3663 9829 7589	58
59	9958 I549 8485 2023	0000 0000 0000 0000	0033 I585 I957 9656	59
60	9984 4674 2037 3248	0039 8409 6248 0833	00I9 4523 0563 5499	60
6I	9997 4220 5025 2453	0000 0000 0000 0000	0006 9450 6849 3385	6I
Remainder terms		$G_{60} = 0{,}824 \times 10^{-I6}$	$K_{92} = 0{,}I24 \times 10^{-28}$	

$n = 36$

No.	Nodes		Weights of Gaussian quadratures		Weights of improved quadratures		No.
	Mantissa	Order	Mantissa	Order	Mantissa	Order	
40	4I5I 324I 0703 4I20	0	6452 2I5I 3076 0235	− 4	6452 2064 2072 3407	− 5	40
4I	4322 4307 0642 7474	0	0000 0000 0000 0000	0	6434 6657 4384 7436	− 5	4I
42	4473 I002 3066 2303	0	6406 43I4 5767 245I	− 4	6406 4556 0530 63I6	− 5	42
43	4642 674I 0I72 707I	0	0000 0000 0000 0000	0	6347 3I60 3763 0765	− 5	43
44	50II 376I 266I 7536	0	6277 3642 5050 4526	− 4	6277 32I6 77I5 0756	− 5	44
45	5I56 5736 7I32 6567	0	0000 0000 0000 0000	0	62I6 7I3I 7432 6702	− 5	45
46	5322 2555 5352 3I37	0	6I25 6370 7502 6342	− 4	6I25 72II 6I40 2457	− 5	46
47	5464 0I3I I266 I4I7	0	0000 0000 0000 0000	0	6024 264I 5538 722I	− 5	47
50	5623 4I47 0342 7725	0	57I2 5273 5036 I776	− 4	57I2 4234 50I5 3644	− 5	50
5I	5760 463I 4573 05I3	0	0000 0000 0000 0000	0	5570 767I 2362 6752	− 5	5I
52	6II3 008I 4764 0705	0	5437 4204 0007 33I6	− 4	5437 55I2 2362 024I	− 5	52
53	6242 3774 0507 5733	0	0000 0000 0000 0000	0	5276 42II III3 I000	− 5	53
54	6366 6546 7I62 002I	0	5I26 I530 6I74 0652	− 4	5I25 7705 6I43 236I	− 5	54
55	6507 634I 4I36 2266	0	0000 0000 0000 0000	0	4746 6720 3200 40I7	− 5	55
56	6625 I452 I745 0476	0	4560 6587 266I 0I72	− 4	456I 0766 2772 4825	− 5	56
57	6736 6325 I402 7346	0	0000 0000 0000 0000	0	4364 3I76 4335 6046	− 5	57
60	7044 3207 72I3 7640	0	4I6I 6577 I734 4062	− 4	4I6I 36I5 6I74 45I7	− 5	60
6I	7I45 6573 2444 3437	0	0000 0000 0000 0000	0	7723 0467 434I 0I53	− 6	6I
62	7242 7465 7732 4I34	0	7267 676I 4662 6407	− 5	7270 6634 3045 4642	− 6	62
63	7333 4346 5340 2453	0	0000 0000 0000 0000	0	6622 5547 5I7I 7050	− 6	63
64	74I7 3676 2757 57I2	0	6I44 074I 4330 53I7	− 5	6I42 62I3 I764 4I56	− 6	64
65	7476 4723 3745 47I7	0	0000 0000 0000 0000	0	5455 0265 3772 5732	− 6	65
66	7550 667I I0I2 0626	0	4756 5746 7462 5000	− 5	4760 5I47 5404 03I2	− 6	66
67	76I6 0503 0I32 7706	0	0000 0000 0000 0000	0	425I 6655 I537 26I3	− 6	67
70	7656 0777 06I3 6755	0	7274 6750 2767 5457	− 6	7266 7403 4443 76II	− 7	70
7I	77I0 7423 2742 6II6	0	0000 0000 0000 0000	0	6033 252I 7552 2747	− 7	7I
72	7736 4326 2234 6527	0	4564 3375 5360 00I5	− 6	4576 7364 I565 02I2	− 7	72
73	7756 6703 5552 04I7	0	0000 0000 0000 0000	0	6624 3I40 I677 2002	−I0	73
74	777I 5064 5I75 4432	0	4050 6405 4267 444I	− 7	7757 3433 2852 4047	−II	74
75	7776 7432 7260 I655	0	0000 0000 0000 0000	0	5540 7620 6482 57II	−I2	75

No.	Nodes	Weights of Gaussian quadratures	Weights of improved quadratures	No.
1	0,0002 4176 8560 3858	0,0000 0000 0000 0000	0,0006 5123 1356 3366	1
2	0014 5625 9090 2615	0037 3541 5789 6244	0018 2356 6363 3887	2
3	0039 2320 4533 6530	0000 0000 0000 0000	0031 0872 0873 9889	3
4	0076 5704 5167 4238	0086 5931 0395 1553	0043 4879 5466 7253	4
5	0126 0581 8741 9837	0000 0000 0000 0000	0055 4582 4726 7962	5
6	0,0187 4803 7453 5252	0,0135 0450 9592 4897	0,0067 4117 6250 1479	6
7	0260 9002 0868 9362	0000 0000 0000 0000	0079 4116 8878 5337	7
8	0346 2150 1051 6759	0182 1613 6956 1927	0091 1544 3522 9866	8
9	0443 0826 9006 2035	0000 0000 0000 0000	0102 5287 5401 4305	9
10	0551 1998 5025 8645	0227 4685 3763 6006	0113 6801 9320 2867	10
11	0,0670 3869 5478 2203	0,0000 0000 0000 0000	0,0124 6546 7690 3240	11
12	0800 3983 9926 8663	0270 5154 1212 4584	0135 2998 7117 9498	12
13	0940 8446 2448 5398	0000 0000 0000 0000	0145 5258 1688 3608	13
14	1091 3342 5791 6875	0310 8739 3280 5142	0155 4024 7181 5419	14
15	1251 5454 4486 6955	0000 0000 0000 0000	0164 9609 3256 1994	15
16	0,1421 1160 7706 5734	0,0348 1429 1617 7052	0,0174 1006 9019 6254	16
17	1599 5815 6074 6246	0000 0000 0000 0000	0182 7497 1293 8278	17
18	1786 4663 8537 8698	0381 9519 3299 3883	0190 9504 3995 4705	18
19	1981 3425 5749 7386	0000 0000 0000 0000	0198 7271 0889 1329	19
20	2183 7541 9296 4254	0411 9649 5880 7946	0206 0053 1267 5948	20
21	0,2393 1659 9040 3364	0,0000 0000 0000 0000	0,0212 7259 8018 4262	21
22	2609 0310 8977 5488	0437 8837 0304 2389	0218 9209 7827 5869	22
23	2830 8415 7896 9484	0000 0000 0000 0000	0224 6132 3475 8653	23
24	3058 0704 9195 8835	0459 4505 6946 8207	0229 7447 2377 0144	24
25	3290 1265 7317 9093	0000 0000 0000 0000	0234 2663 3650 2818	25
26	0,3526 4096 5009 1492	0,0476 4512 1456 1598	0,0238 2071 8208 0161	26
27	3766 3568 3820 4432	0000 0000 0000 0000	0241 5925 0816 3271	27
28	4009 3940 0332 2147	0488 7166 7693 1644	0244 3760 8080 8283	28
29	4254 8936 3174 1391	0000 0000 0000 0000	0246 5167 5016 3235	29
30	4502 2234 3923 8292	0496 1250 5613 3362	0248 0451 7412 1383	30
31	0,4750 7909 6083 4657	0,0000 0000 0000 0000	0,0248 9915 0258 0998	31
32	5000 0000 0000 0000	0498 6027 2396 7132	0249 3185 8596 8880	32

No.	Nodes		Weights of Gaussian quadratures		Weights of improved quadratures		No.
	Mantissa	Order	Mantissa	Order	Mantissa	Order	
I	7730 I500 7537 7606	−I4	0000 0000 0000 0000	0	5253 3545 4052 4I32	−I2	I
2	5755 7762 258I 40I5	−II	75I4 6740 5I64 2444	−I0	7360 2273 6222 II33	−II	2
3	40I0 707I 4527 I2I5	− 7	0000 0000 0000 0000	0	6273 5656 3707 276I	−I0	3
4	7656 3763 2237 7205	− 7	4335 7707 57I7 7653	− 6	4350 0I27 II63 6765	− 7	4
5	6350 4242 636I 6247	− 6	0000 0000 0000 0000	0	5533 26I6 I307 I775	− 7	5
6	463I 2573 7564 2060	− 5	6724 I004 606I 3267	− 6	67I6 2425 6077 3I45	− 7	6
7	6533 5275 2304 50I3	− 5	0000 0000 0000 0000	0	404I 5655 I200 245I	− 6	7
I0	4334 7506 330I 3006	− 4	4523 5002 0370 77I7	− 5	4525 4360 7507 3374	− 6	I0
II	5527 6226 3I22 745I	− 4	0000 0000 0000 0000	0	5I77 5655 I050 0272	− 6	II
I2	7034 2576 24I2 6I56	− 4	5645 3634 I000 6305	− 5	5644 0355 63I5 5776	− 6	I2
I3	4224 5625 3I22 4746	− 3	0000 0000 0000 0000	0	6303 5766 0074 5530	− 6	I3
I4	5076 5755 3556 8582	− 3	673I 546I 4640 5602	− 5	6732 634I 0277 303I	− 6	I4
I5	60I2 7582 6243 47I2	− 3	0000 0000 0000 0000	0	7346 6763 4662 0322	− 6	I5
I6	6770 0530 3526 3752	− 3	7752 5375 I352 2567	− 5	775I 6205 2652 3626	− 6	I6
I7	4002 4203 234I 0205	− 2	0000 0000 0000 0000	0	4I62 I320 5030 I060	− 5	I7
20	4430 2664 4336 6506	− 2	435I 4556 I756 3470	− 4	435I 76I7 5036 7277	− 5	20
2I	5074 6022 II04 324I	− 2	0000 0000 0000 0000	0	4533 2544 4025 5302	− 5	2I
22	5556 7444 7603 6757	− 2	4707 I220 I025 6505	− 4	4706 6465 5363 6500	− 5	22
23	6256 I664 6452 3I23	− 2	0000 0000 0000 0000	0	5054 6030 3264 04II	− 5	23
24	677I 67I6 24I0 I604	− 2	52I3 6650 I204 4364	− 4	52I4 II62 0046 I254	− 5	24
25	7520 755I 062I I240	− 2	0000 0000 0000 0000	0	5344 I737 0624 4I65	− 5	25
26	4I3I 2427 50I7 3562	− I	5465 5557 II46 6763	− 4	5465 34I6 4I53 5855	− 5	26
27	44I7 0I50 I005 6676	− I	0000 0000 0000 0000	0	5600 03I7 I60I 7I66	− 5	27
30	47II I275 6545 6573	− I	5703 0342 2377 6702	− 4	5703 2365 740I 0567	− 5	30
3I	5207 2I30 6544 64I3	− I	0000 0000 0000 0000	0	5776 4466 I304 2I04	− 5	3I
32	55I0 6533 2I76 I553	− I	6062 3607 7I67 66I7	− 4	6062 I652 5527 30I5	− 5	32
33	60I5 3I44 3420 3742	− I	0000 0000 0000 0000	0	6I36 4637 0II5 642I	− 5	33
34	6324 3755 6603 2635	− I	6202 6650 3I6I 3405	− 4	6203 0540 7377 635I	− 5	34
35	6635 4675 7I64 076I	− I	0000 0000 0000 0000	0	6237 III7 I77I I5I2	− 5	35
36	7I50 I6I3 0I42 464I	− I	6263 3I73 4424 756I	− 4	6263 I327 7050 0225	− 5	36
37	7463 662I I626 I052	− I	0000 0000 0000 0000	0	6277 45I5 4035 0270	− 5	37
40	4000 0000 0000 0000	0	6303 5II0 7520 2075	− 4	6303 6745 5232 54I5	− 5	40

No.	Nodes	Weights of Gaussian quadratures	Weights of improved quadratures	No.
33	0,5249 2090 3916 5343	0,0000 0000 0000 0000	0,0248 9915 0258 0998	33
34	0,5497 7765 6076 1708	0,0496 1250 5618 8362	0,0248 0451 7412 1383	34
35	5745 1063 6825 8609	0000 0000 0000 0000	0246 5167 5016 3235	35
36	5990 6059 9667 7853	0488 7166 7693 1644	0244 3760 8080 8283	36
37	6233 6431 6179 5567	0000 0000 0000 0000	0241 5925 0816 3271	37
38	6473 5903 4990 8508	0476 4512 1456 1598	0238 2071 8208 0161	38
39	0,6709 8734 2682 0907	0000 0000 0000 0000	0234 2663 3650 2818	39
40	6941 9295 0804 1165	0459 4505 6946 8207	0229 7447 2377 0144	40
41	7169 1584 2103 0516	0000 0000 0000 0000	0224 6132 3475 8653	41
42	7390 9689 1022 4512	0437 8837 0304 2389	0218 9209 7827 5869	42
43	7606 8340 0959 6636	0000 0000 0000 0000	0212 7259 8018 4262	43
44	0,7816 2458 0703 5746	0,0411 9649 5880 7946	0,0206 0053 1267 5948	44
45	8018 6574 4250 2614	0000 0000 0000 0000	0198 7271 0889 1329	45
46	8213 5336 1462 1302	0881 9519 3299 3883	0190 9504 3995 4705	46
47	8400 4184 3925 3754	0000 0000 0000 0000	0182 7497 1293 8278	47
48	8578 8839 2293 4266	0348 1429 1617 7052	0174 1006 9019 6254	48
49	0,8748 4545 5513 3045	0,0000 0000 0000 0000	0,0164 9609 3256 1994	49
50	8908 6657 4208 3125	0310 8739 3280 5142	0155 4024 7181 5419	50
51	9059 1553 7551 4602	0000 0000 0000 0000	0145 5258 1688 3608	51
52	9199 6016 0073 1337	0270 5154 1212 4584	0135 2998 7117 9498	52
53	9329 6130 4521 7797	0000 0000 0000 0000	0124 6546 7690 3240	53
54	0,9448 8001 4974 1355	0,0227 4685 3763 6006	0,0113 6801 9320 2867	54
55	9556 9173 0993 7965	0000 0000 0000 0000	0102 5287 5401 4305	55
56	9653 7849 8948 3241	0182 1613 6956 1927	0091 1544 3522 9866	56
57	9739 0997 9131 0638	0000 0000 0000 0000	0079 4116 8878 5337	57
58	9812 5196 2546 4748	0135 0450 9592 4897	0067 4117 6250 1479	58
59	0,9873 9418 1258 0163	0,0000 0000 0000 0000	0,0055 4532 4726 7962	59
60	9923 4295 4832 5762	0086 5931 0395 1553	0043 4879 5466 7253	60
61	9960 7679 5466 3470	0000 0000 0000 0000	0031 0872 0873 9889	61
62	9985 4374 0909 7385	0037 3541 5789 6244	0018 2356 6363 3887	62
63	9997 5823 1439 6142	0000 0000 0000 0000	0006 5123 1356 3366	63
	Remainder terms	$G_{62} = 0{,}213 \times 10^{-16}$	$K_{96} = 0{,}150 \times 10^{-29}$	

No.	Nodes		Weights of Gaussian quadratures		Weights of improved quadratures		No.
	Mantissa	Order	Mantissa	Order	Mantissa	Order	
4I	4I46 0467 3064 7352	0	0000 0000 0000 0000	0	6277 45I5 4035 0270	− 5	4I
42	43I3 7072 37I6 5457	0	6263 3I73 4424 756I	− 4	6263 I327 7050 0225	− 5	42
43	446I I44I 0305 7407	0	0000 0000 0000 0000	0	6237 III7 I77I I5I2	− 5	43
44	4625 60II 0476 246I	0	6202 6650 3I6I 3405	− 4	6203 0540 7377 635I	− 5	44
45	477I 23I5 6I67 60I6	0	0000 0000 0000 0000	0	6I36 4637 0II5 642I	− 5	45
46	5I33 4522 2700 7II2	0	6062 3607 7I67 66I7	− 4	6062 I652 5527 30I5	− 5	46
47	5274 2723 45I5 4572	0	0000 0000 0000 0000	0	5776 4466 I304 2I04	− 5	47
50	5433 324I 05I5 0502	0	5703 0342 2377 6702	− 4	5703 2365 740I 0567	− 5	50
5I	5570 37I3 7375 0440	0	0000 0000 0000 0000	0	5600 03I7 I60I 7I66	− 5	5I
52	5723 2564 I370 2I06	0	5465 5557 II46 6763	− 4	5465 34I6 4I53 5355	− 5	52
53	6053 6045 5633 5527	0	0000 0000 0000 0000	0	5344 I737 0624 4I65	− 5	53
54	620I 42I4 3275 7436	0	52I3 6650 I204 4364	− 4	52I4 II62 0046 I254	− 5	54
55	6324 3422 6265 3I53	0	0000 0000 0000 0000	0	5054 6030 3264 04II	− 5	55
56	6444 2066 6037 0204	0	4707 I220 I025 6505	− 4	4706 6465 5363 6500	− 5	56
57	6560 6373 3556 7I27	0	0000 0000 0000 0000	0	4533 2544 4025 5302	− 5	57
60	667I 7222 67I0 2256	0	435I 4556 I756 3470	− 4	435I 76I7 5036 7277	− 5	60
6I	6777 2737 I307 5736	0	0000 0000 0000 0000	0	4I62 I320 5030 I060	− 5	6I
62	7I00 7724 7425 I402	0	7752 5375 I352 2567	− 5	775I 6205 2652 3626	− 6	62
63	7I76 5024 5I53 4306	0	0000 0000 0000 0000	0	7346 6763 4662 0322	− 6	63
64	7270 I202 2422 I424	0	673I 546I 4640 5602	− 5	6732 634I 0277 303I	− 6	64
65	7355 32I5 2465 5303	0	0000 0000 0000 0000	0	6303 5766 0074 5530	− 6	65
66	7436 I650 0657 247I	0	5645 3634 I000 6305	− 5	5644 0355 63I5 5776	− 6	66
67	75I2 4066 4632 64I5	0	0000 0000 0000 0000	0	5I77 5655 I050 0272	− 6	67
70	7562 I4I3 4623 7237	0	4523 5002 0370 77I7	− 5	4525 4360 7507 3374	− 6	70
7I	7625 I052 053I 6657	0	0000 0000 0000 0000	0	404I 5655 I200 245I	− 6	7I
72	7663 I524 I004 2736	0	6724 I004 606I 3267	− 6	67I6 2425 6077 3I45	− 7	72
73	77I4 2735 35I4 I6I5	0	0000 0000 0000 0000	0	5533 26I6 I307 I775	− 7	73
74	7740 5060 0626 6002	0	4335 7707 57I7 7653	− 6	4850 0I27 II63 6765	− 7	74
75	7757 7343 43I5 2432	0	0000 0000 0000 0000	0	6273 5656 3707 276I	−I0	75
76	7772 0220 0I55 2463	0	75I4 6740 5I64 2444	−I0	7360 2273 6222 II33	−II	76
77	7777 0047 6277 0240	0	0000 0000 0000 0000	0	5253 3545 4052 4I32	−I2	77

No.	Nodes	Weights of Gaussian quadratures	Weights of improved quadratures	No.
I	0,0002 2704 8937 8I78	0,0000 0000 0000 0000	0,0006 II68 0408 9757	I
2	00I3 6806 9075 2592	0035 0930 5004 7350	00I7 I340 9387 8862	2
3	0036 8598 2368 5I40	0000 0000 0000 0000	0029 2086 8539 5833	3
4	007I 9424 4227 3658	008I 37I9 7365 4528	0040 8625 20I9 2658	4
5	0II8 4485 8I92 668I	0000 0000 0000 0000	0052 II99 3699 4034	5
6	0,0I76 I887 2206 2468	0,0I26 9603 2654 68I0	0,0063 3802 7403 3272	6
7	0245 2265 7575 6694	0000 0000 0000 0000	0074 6805 I803 0430	7
8	0325 4696 2081 I302	0I7I 3693 I456 5I07	0085 7490 2604 892I	8
9	04I6 6I36 6674 3I78	0000 0000 0000 0000	0096 4938 57I5 I634	9
I0	05I8 3942 2II6 9739	02I4 I794 90II II33	0I07 0445 6592 4II0	I0
II	0,0630 65II 5527 3447	0000 0000 0000 0000	0,0II7 4332 9836 08I7	II
I2	0753 I6I9 3I33 7I50	0254 9902 963I I88I	0I27 5284 7740 4473	I2
I3	0885 5852 493I 9743	0000 0000 0000 0000	0I37 2604 92II I052	I3
I4	I027 58I0 20I6 0288	0293 4204 6739 2678	0I46 6847 8344 8I03	I4
I5	II78 8587 400I 098I	0000 0000 0000 0000	0I55 8I66 2780 9869	I5
I6	0,I339 0894 0629 8552	0,0329 IIII I388 I809	0,0I64 5753 882I 95I8	I6
I7	I507 8672 II02 3948	0000 0000 0000 0000	0I72 9I06 I372 3665	I7
I8	I684 7786 6534 8924	036I 7289 7054 4243	0I80 8488 4737 82II	I8
I9	I869 4353 II49 0880	0000 0000 0000 0000	0I88 3956 5322 8067	I9
20	206I 42I2 I379 6I88	0390 9694 7893 5352	0I95 497I 0066 6533	20
2I	0,2260 2684 2900 4238	0,0000 0000 0000 0000	0,0202 II74 6I85 I865	2I
22	2465 5004 5533 8853	04I6 5596 2II3 4734	0208 2700 9992 82I5	22
23	2676 6534 5759 0039	0000 0000 0000 0000	02I3 9555 7798 2234	23
24	2893 2436 I934 6823	0488 2604 6502 20I9	02I9 I377 20I5 0699	24
25	3II4 7528 9422 9394	0000 0000 0000 0000	0223 793I 9374 8835	25
26	0,3340 6569 8858 9362	0,0455 8693 9347 88I9	0,0227 929I 3282 2735	26
27	3570 4377 0705 270I	0000 0000 0000 0000	023I 5437 8369 0I29	27
28	3803 563I 8873 93I5	0469 22I9 9540 4023	0234 6I48 4I40 85I8	28
29	4039 48I9 5508 4288	0000 0000 0000 0000	0237 I303 0936 94I2	29
30	4277 640I 9208 60I8	0478 I936 0039 6374	0239 0945 4368 4942	30
3I	0,45I7 4865 I56I 5528	0,0000 0000 0000 0000	0,0240 5048 4592 7289	3I
32	4758 46I6 7I56 I308	0482 7004 4257 3639	024I 3509 6537 8887	32
33	5000 0000 0000 0000	0000 0000 0000 0000	024I 63I9 I993 2839	33

No.	Nodes		Weights of Gaussian quadratures		Weights of improved quadratures		No.
	Mantissa	Order	Mantissa	Order	Mantissa	Order	
I	734I 1774 I464 3357	-I4	0000 0000 0000 0000	0	5005 4455 32II 5475	-I2	I
2	5465 0307 6773 II20	-II	7I37 6I36 I64I 7I42	-I0	70II 2I72 4443 7052	-II	2
3	743I 0205 5637 254I	-I0	0000 0000 0000 0000	0	5766 60I2 6647 5034	-I0	3
4	7273 666I 6550 3II0	- 7	4I25 074I II03 2266	- 6	4I36 2767 3III 5I34	- 7	4
5	604I 0357 5I00 622I	- 6	0000 0000 0000 0000	0	5254 4537 I743 0724	- 7	5
6	4405 2563 7726 I052	- 5	6400 I405 205I 4377	- 6	6372 7472 I530 7603	- 7	6
7	62I6 I675 4280 6536	- 5	0000 0000 0000 0000	0	75I3 32I7 I022 I700	- 7	7
I0	4I24 7766 4602 0057	- 4	4306 I300 0I2I 5256	- 5	4307 6677 4404 40II	- 6	I0
II	5252 2433 7537 3473	- 4	0000 0000 0000 0000	0	474I 4I65 0II2 6760	- 6	II
I2	6505 2622 7242 632I	- 4	5367 226I 643I 5244	- 5	5366 0676 576I 4I06	- 6	I2
I3	4022 4II0 4I32 532I	- 3	0000 0000 0000 0000	0	6006 3430 3I36 2450	- 6	I3
I4	4643 7540 2457 I275	- 3	64I6 I527 2260 7737	- 5	64I7 052I 7747 4075	- 6	I4
I5	5525 7054 0056 5454	- 3	0000 0000 0000 0000	0	70I6 I47I 0734 IIII	- 6	I5
I6	6447 I326 7707 3207	- 3	7405 7273 3III III4	- 5	7405 20I6 5427 46I7	- 6	I6
I7	7426 7046 I277 2763	- 3	0000 0000 0000 0000	0	7764 5073 0077 0744	- 6	I7
20	422I 7554 705I 3304	- 2	4I54 67I5 I426 2540	- 4	4I55 0765 7006 4763	- 5	20
2I	4646 3725 4363 4462	- 2	0000 0000 0000 0000	0	4332 2773 7067 0054	- 5	2I
22	53I0 2566 I667 4007	- 2	4502 50I0 073I 5360	- 4	4502 3300 4400 3443	- 5	22
23	5766 7040 0036 7665	- 2	0000 0000 0000 0000	0	4645 2556 4735 2574	- 5	23
24	646I 3353 455I 6465	- 2	5002 2037 0223 3750	- 4	5002 3266 5303 6752	- 5	24
25	7I67 I624 522I 50I7	- 2	0000 0000 0000 0000	0	5I3I I432 4670 4242	- 5	25
26	7707 3635 363I 2326	- 2	525I 756I I365 7536	- 4	525I 6547 6625 650I	- 5	26
27	4220 5556 5III 6552	- I	0000 0000 0000 0000	0	5364 2674 50I3 63I5	- 5	27
30	4502 II22 50I6 2034	- I	5470 I360 6I60 4766	- 4	5470 2202 6I06 00I5	- 5	30
3I	4767 4660 3262 5325	- I	0000 0000 0000 0000	0	5565 2325 4644 6265	- 5	3I
32	5260 5250 6I52 04I0	- I	5653 4536 6637 5562	- 4	5653 4064 II0I 3776	- 5	32
33	5554 7I60 73I5 3335	- I	0000 0000 0000 0000	0	5732 7I00 I665 2363	- 5	33
34	6053 7020 I7I4 3755	- I	6003 0576 0I73 3767	- 4	6003 III4 3063 3765	- 5	34
35	6355 III4 2037 45I4	- I	0000 0000 0000 0000	0	6044 0724 427I 7522	- 5	35
36	6660 I742 5435 0033	- I	6075 7073 5574 3066	- 4	6075 6702 4447 70I5	- 5	36
37	7I64 563I 32I7 0034	- I	0000 0000 0000 0000	0	6I20 2605 46I6 55I3	- 5	37
40	7472 I033 6673 4685	- I	6I33 33I7 2566 I675	- 4	6I33 3367 224I 6364	- 5	40
4I	4000 0000 0000 0000	0	0000 0000 0000 0000	0	6I37 0742 730I 4626	- 5	4I

No.	Nodes	Weights of Gaussian quadratures	Weights of improved quadratures	No.
34	524I 5383 2843 8692	0482 7004 4257 3639	024I 3509 6537 8887	34
35	5482 5I34 8438 4472	0000 0000 0000 0000	0240 5048 4592 7289	35
36	0,5722 3598 079I 3982	0,0478 I936 0039 6374	0,0239 0945 4368 4942	36
37	5960 5I80 449I 57I2	0000 0000 0000 0000	0237 I308 0936 94I2	37
38	6I96 4368 II26 0685	0469 22I9 9540 4023	0234 6I48 4I40 85I8	38
39	6429 5622 9294 7299	0000 0000 0000 0000	023I 5437 8369 0I29	39
40	6659 3430 II4I 0638	0455 8693 9347 88I9	0227 929I 3282 2735	40
4I	0,6885 247I 0577 0606	0,0000 0000 0000 0000	0,0223 793I 9374 8835	4I
42	7I06 7563 8065 3I77	0438 2604 6502 20I9	02I9 I377 20I5 0699	42
43	7323 3465 4240 996I	0000 0000 0000 0000	02I3 9555 7798 2234	43
44	7534 4995 4466 II47	04I6 5596 2II3 4734	0208 2700 9992 82I5	44
45	7739 73I5 7099 5762	0000 0000 0000 0000	0202 II74 6I85 I865	45
46	0,7938 5787 8620 38I2	0,0390 9694 7893 5852	0,0I95 497I 0066 6533	46
47	8I30 5646 8850 9I20	0000 0000 0000 0000	0I88 3956 5322 8067	47
48	83I5 22I3 3465 I076	036I 7289 7054 4243	0I80 8488 4737 82II	48
49	8492 I327 8897 6052	0000 0000 0000 0000	0I72 9I06 I372 3665	49
50	8660 9I05 9370 I448	0329 IIII I388 I809	0I64 5753 882I 95I8	50
5I	0,882I I4I2 5998 90I9	0,0000 0000 0000 0000	0,0I55 8I66 2780 9869	5I
52	8972 4I89 7983 97I2	0293 4204 6739 2678	0I46 6847 8344 8I03	52
53	9II4 4I47 5068 0257	0000 0000 0000 0000	0I37 2604 92II I052	53
54	9246 8380 6866 2850	0254 9902 963I I88I	0I27 5284 7740 4473	54
55	9369 3488 4472 6553	0000 0000 0000 0000	0II7 4332 9836 08I7	55
56	0,948I 6057 7883 026I	0,02I4 I794 90II II33	0,0I07 0445 6592 4II0	56
57	9583 3863 3325 6822	0000 0000 0000 0000	0096 4938 57I5 I634	57
58	9674 5303 7968 8698	0I7I 3693 I456 5I07	0085 7490 2604 892I	58
59	9754 7734 2424 3306	0000 0000 0000 0000	0074 6805 I803 0430	59
60	9823 8II2 7793 7582	0I26 9603 2654 63I0	0063 3802 7403 3272	60
6I	0,988I 55I4 I807 33I9	0,0000 0000 0000 0000	0,0052 II99 3699 4034	6I
62	9928 0575 5772 6342	008I 37I9 7365 4528	0040 8625 20I9 2658	62
63	9963 I40I 763I 4860	0000 0000 0000 0000	0029 2086 8539 5833	63
64	9986 3I93 0924 7408	0035 0930 5004 7350	00I7 I340 9387 8862	64
65	9997 7295 I062 I822	0000 0000 0000 0000	0006 II68 0408 9757	65
Remainder terms		$G_{64} = 0,549 \times 10^{-17}$	$K_{98} = 0,176 \times 10^{-30}$	

$n = 40$

No.	Nodes		Weights of Gaussian quadratures		Weights of improved quadratures		No.
	Mantissa	Order	Mantissa	Order	Mantissa	Order	
42	4I42 7362 0442 I46I	0	6I33 83I7 2566 I675	- 4	6I33 8367 224I 6364	- 5	42
43	4305 5063 2270 376I	0	0000 0000 0000 0000	0	6I20 2605 46I6 55I3	- 5	43
44	4447 70I6 5I6I 3762	0	6075 7073 5574 3066	- 4	6075 6702 4447 70I5	- 5	44
45	46II 333I 6760 I53I	0	0000 0000 0000 0000	0	6044 0724 427I 7522	- 5	45
46	4752 0367 708I 60II	0	6003 0576 0I73 3767	- 4	6003 III4 3063 3765	- 5	46
47	5III 4307 428I 222I	0	0000 0000 0000 0000	0	5732 7I00 I665 2363	- 5	47
50	5247 5253 47I2 7573	0	5653 4536 6637 5562	- 4	5653 4064 II0I 3776	- 5	50
5I	5404 I447 6246 5225	0	0000 0000 0000 0000	0	5565 2325 4644 6265	- 5	5I
52	5536 7326 5370 676I	0	5470 I360 6I60 4766	- 4	5470 2202 6I06 00I5	- 5	52
53	5667 5II0 5333 05I2	0	0000 0000 0000 0000	0	5364 2674 50I3 63I5	- 5	53
54	60I6 I030 508I 53I2	0	525I 756I I365 7536	- 4	525I 6547 6625 650I	- 5	54
55	6I42 I432 6533 4574	0	0000 0000 0000 0000	0	5I3I I432 4670 4242	- 5	55
56	6263 5I05 0645 4262	0	5002 2037 0223 3750	- 4	5002 3266 5303 6752	- 5	56
57	6402 2I67 7770 2022	0	0000 0000 0000 0000	0	4645 2556 4735 2574	- 5	57
60	65I5 7242 3422 0776	0	4502 50I0 073I 5360	- 4	4502 3300 4400 3443	- 5	60
6I	6626 30I2 4703 0663	0	0000 0000 0000 0000	0	4332 2773 7067 0054	- 5	6I
62	6733 4044 6I65 5II6	0	4I54 67I5 I426 2540	- 4	4I55 0765 7006 4763	- 5	62
63	7085 I073 I650 050I	0	0000 0000 0000 0000	0	7764 5073 0077 0744	- 6	63
64	7I33 0645 I007 0457	0	7405 7273 3III III4	- 5	7405 20I6 5427 46I7	- 6	64
65	7225 2072 3772 I232	0	0000 0000 0000 0000	0	70I6 I47I 0734 IIII	- 6	65
66	73I3 4023 7532 0650	0	64I6 I527 2260 7737	- 5	64I7 052I 7747 4075	- 6	66
67	7375 5366 7364 5245	0	0000 0000 0000 0000	0	6006 3430 3I36 2450	- 6	67
70	7453 5246 6425 6462	0	5367 226I 643I 5244	- 5	5366 0676 576I 4I06	- 6	70
7I	7525 2656 20I2 02I4	0	0000 0000 0000 0000	0	474I 4I65 0II2 6760	- 6	7I
72	7572 5400 4547 6775	0	4306 I300 0I2I 5256	- 5	4307 6677 4404 40II	- 6	72
73	7633 4342 0473 I625	0	0000 0000 0000 0000	0	75I3 82I7 I022 I700	- 7	73
74	7667 6524 300I 2356	0	6400 I405 205I 4377	- 6	6372 7472 I530 7603	- 7	74
75	77I7 3674 2026 77I5	0	0000 0000 0000 0000	0	5254 4537 I743 0724	- 7	75
76	7742 4204 4705 I363	0	4I25 074I II03 2266	- 6	4I36 2767 3III 5I34	- 7	76
77	7760 7I57 3644 30I2	0	0000 0000 0000 0000	0	5766 60I2 6647 5034	-I0	77
I00	7772 3I27 470I 0046	0	7I37 6I36 I64I 7I42	-I0	70II 2I72 4443 7052	-II	I00
I0I	7777 0436 6003 68I3	0	0000 0000 0000 0000	0	5005 4455 82II 5475	-I2	I0I

No.	Nodes	Weights of Gaussian quadratures	Weights of improved quadratures	No.
I	0,0002 I375 9799 0I47	0,0000 0000 0000 0000	0,0005 7580 0645 024I	I
2	00I2 8765 2876 7724	0033 03II 3928 7987	00I6 I254 4764 3I47	2
3	0034 6982 2283 6973	0000 0000 0000 0000	0027 4947 5I50 3482	3
4	0067 72I3 6884 6788	0076 6085 0756 4673	0038 4732 7993 77I2	4
5	0III 5III 6802 2I99	0000 0000 0000 0000	0049 078I 74I0 643I	5
6	0,0165 8854 5I55 0036	0,0II9 5777 4050 8747	0,0059 69II 6927 2883	6
7	0230 9I58 4225 8805	0000 0000 0000 0000	0070 3572 698I 0250	7
8	0306 528I 3694 4I58	0I6I 50I7 93I6 I645	0080 8I58 I326 6994	8
9	0392 4429 994I I673	0000 0000 0000 0000	0090 97I8 8976 2772	9
I0	0488 4I6I 6I28 2882	0202 0077 0665 8348	0I00 9563 7448 03I5	I0
II	0,0594 3I52 I3I6 9289	0,0000 0000 0000 0000	0,0II0 8II7 I430 6476	II
I2	0709 95I7 366I 7480	0240 7387 I409 3558	0I20 4062 6326 77I0	I2
I3	0835 0I33 0456 66I4	0000 0000 0000 0000	0I29 6635 9709 6I87	I3
I4	0969 I882 I862 9I67	0277 3542 33I5 83I8	0I38 6470 38I6 6680	I4
I5	III2 2272 938I 3748	0000 0000 0000 0000	0I47 385I 3784 938I	I5
I6	0,I263 8475 I775 2I89	0,03II 5324 I265 I587	0,0I55 79I5 8804 I789	I6
I7	I423 6780 9I38 8906	0000 0000 0000 0000	0I63 8053 0538 6839	I7
I8	I59I 3402 00I5 I286	0342 9728 6409 3284	0I7I 4643 8667 2789	I8
I9	I766 4970 85I2 4II0	0000 0000 0000 0000	0I78 7903 I777 5430	I9
20	I948 7882 708I 8I05	037I 3992 742I 977I	0I85 7I92 3996 9665	20
2I	0,2I87 7853 4I54 47I8	0,0000 0000 0000 0000	0,0I92 2009 58I0 302I	2I
22	2333 0504 7606 8262	0396 56I8 2397 4434	0I98 268I I284 07I0	22
23	2534 I79I 3364 5659	0000 0000 0000 0000	0203 9248 6444 2664	23
24	2740 7499 I363 7747	04I8 2393 8033 5I94	0209 I36I 4002 43I3	24
25	2952 2862 7464 29I0	0000 0000 0000 0000	02I3 8549 4002 5390	25
26	0,3I68 3037 II25 9633	0,0436 24I4 3809 4222	0,02I8 I052 6I50 2875	26
27	3388 348I 5I5I 455I	0000 0000 0000 0000	022I 9069 35I7 7029	27
28	36II 9545 I423 75I5	0450 4097 9330 3I93	0225 2I96 3737 2I32	28
29	3838 6II7 230I 3794	0000 0000 0000 0000	0228 0078 8504 0I55	29
30	4067 8035 0586 0042	0460 6I99 332I 6584	0230 2957 0580 83I4	30
3I	0,4299 0446 9829 II25	0,0000 0000 0000 0000	0,0232 I057 I427 6354	3I
32	458I 8446 7072 6333	0466 782I 3032 798I	0233 4050 4597 5420	32
33	4765 6736 3309 3355	0000 0000 0000 0000	0234 I634 33I6 8025	33
34	0,5000 0000 0000 0000	0,0468 8422 3080 I050	0,0234 4072 2395 2853	34

No.	Nodes Mantissa	Order	Weights of Gaussian quadratures Mantissa	Order	Weights of improved quadratures Mantissa	Order	No.
I	7002 2265 4221 2536	-14	0000 0000 0000 0000	0	4557 0523 6037 0215	-12	I
2	5214 3165 7721 4726	-11	6607 4416 2161 2171	-10	6465 6006 0325 4277	-11	2
3	7065 6621 4207 1667	-10	0000 0000 0000 0000	0	5503 0211 6641 3313	-10	3
4	6736 4315 1614 6052	- 7	7660 3737 5653 1072	- 7	7702 1563 7220 2376	-10	4
5	5553 1454 3752 7251	- 6	0000 0000 0000 0000	0	5015 0701 5060 0165	- 7	5
6	4176 2263 2733 5011	- 5	6076 5212 0755 0120	- 6	6071 4224 7750 6724	- 7	6
7	5722 5165 0240 7637	- 5	.0000 0000 0000 0000	0	7150 5764 4224 1147	- 7	7
I0	7661 5634 0262 0770	- 5	4104 6541 3771 4512	- 5	4106 4233 6776 0577	- 6	I0
II	5013 7241 4315 1741	- 4	0000 0000 0000 0000	0	4520 6140 2215 7605	- 6	II
I2	6200 7045 3764 3265	- 4	5127 6026 1272 1745	- 5	5126 4054 1257 5371	- 6	I2
I3	7466 7167 4150 3000	- 4	0000 0000 0000 0000	0	5530 6715 1625 4023	- 6	I3
I4	4426 2752 7206 4214	- 3	6123 3221 2310 2570	- 5	6124 3014 0457 7546	- 6	I4
I5	5260 1276 6675 7354	- 3	0000 0000 0000 0000	0	6507 0332 5632 4673	- 6	I5
I6	6147 6540 0442 2405	- 3	7063 2546 0052 0576	- 5	7062 4810 2730 3060	- 6	I6
I7	7074 4276 0241 7160	- 3	0000 0000 0000 0000	0	7427 4716 5312 5510	- 6	I7
20	4026 5401 0745 4677	- 2	7763 2425 0313 7200	- 5	7763 7671 6651 5030	- 6	20
2I	4434 4335 7420 2321	- 2	0000 0000 0000 0000	0	4143 0166 2750 6607	- 5	2I
22	5057 2007 2351 2575	- 2	4307 5517 5570 1606	- 4	4307 3260 1243 1740	- 5	22
23	5516 1651 2072 1214	- 2	0000 0000 0000 0000	0	4447 3414 0611 4437	- 5	23
24	6170 7120 5477 7462	- 2	4602 0011 2647 1607	- 4	4602 2045 6141 5307	- 5	24
25	6656 4302 4461 7400	- 2	0000 0000 0000 0000	0	4727 1566 0621 2174	- 5	25
26	7356 3604 5535 5047	- 2	5046 7205 3173 1436	- 4	5046 5311 6441 3703	- 5	26
27	4033 7776 1103 7044	- I	0000 0000 0000 0000	0	5160 7044 6271 3607	- 5	27
30	4305 1616 5221 7047	- I	5264 7623 7013 1714	- 4	5265 1407 0134 7677	- 5	30
3I	4562 4064 7171 5254	- I	0000 0000 0000 0000	0	5363 0241 5263 0321	- 5	3I
32	5043 3627 III3 5452	- I	5452 7472 7372 1666	- 4	5452 5775 0414 2352	- 5	32
33	5327 5701 6072 2266	- I	0000 0000 0000 0000	0	5534 4501 6736 2027	- 5	33
34	5616 7234 1552 4156	- I	5607 6343 6464 0654	- 4	5607 7773 1542 6512	- 5	34
35	6II0 4563 4647 5057	- I	0000 0000 0000 0000	0	5654 4270 0745 2014	- 5	35
36	6404 2603 4715 4070	- I	5712 5600 1347 5503	- 4	5712 4202 4370 4775	- 5	36
37	6701 6160 2312 1770	- I	0000 0000 0000 0000	0	5742 2030 5150 0126	- 5	37
40	7200 3713 3044 1651	- I	5763 0647 3151 2736	- 4	5763 2225 7726 2403	- 5	40
4I	7500 0243 1370 II24	- I	0000 0000 0000 0000	0	5775 1641 4571 5067	- 5	4I
42	4000 0000 0000 0000	0	6000 4653 6323 6434	- 4	6000 3302 0052 1274	- 5	42

No.	Nodes	Weights of Gaussian quadratures	Weights of improved quadratures	No.
35	5234 8263 6690 6645	0000 0000 0000 0000	0234 I634 33I6 8025	35
36	5468 I553 2927 3667	0466 782I 3032 798I	0233 4050 4597 5420	36
37	5700 9553 0I70 8875	0000 0000 0000 0000	0232 I057 I427 6854	37
38	0,5932 I964 94I3 9958	0,0460 6I99 332I 6584	0,0230 2957 0580 83I4	38
39	6I6I 3882 7698 6206	0000 0000 0000 0000	0228 0078 8504 0I55	39
40	6388 0454 8576 2485	0450 4097 9330 3I93	0225 2I96 3737 2I32	40
4I	66II 65I8 4848 5449	0000 0000 0000 0000	022I 9069 35I7 7029	4I
42	683I 6962 8874 0367	0436 24I4 3809 4222	02I8 I052 6I50 2875	42
43	0,7047 7I37 2535 7090	0,0000 0000 0000 0000	0,02I3 8549 4002 5390	43
44	7259 2500 8636 2253	04I8 2398 8033 5I94	0209 I36I 4002 48I3	44
45	7465 8208 6635 434I	0000 0000 0000 0000	0203 9248 6444 2664	45
46	7666 9495 2393 I738	0396 56I8 2897 4434	0I98 263I I284 07I0	46
47	7862 2I46 5845 5282	0000 0000 0000 0000	0I92 2009 53I0 302I	47
48	0,805I 2II7 29I8 I895	0,037I 3992 742I 977I	0,0I85 7I92 3996 9665	48
49	8233 5029 I487 5890	0000 0000 0000 0000	0I78 7903 I777 5430	49
50	8408 6597 9984 87I4	0342 9728 6409 3284	0I7I 4643 8667 2789	50
5I	8576 32I9 086I I094	0000 0000 0000 0000	0I63 8053 0538 6839	5I
52	8736 I524 8224 78II	03II 5324 I265 I587	0I55 79I5 8304 I789	52
53	0,8887 7727 06I8 6252	0,0000 0000 0000 0000	0,0I47 385I 3784 938I	53
54	9030 8II7 8I37 0833	0277 3542 33I5 83I8	0I38 6470 38I6 6680	54
55	9I64 9866 9543 3386	0000 0000 0000 0000	0I29 6635 9709 6I87	55
56	9290 0482 6338 2520	0240 7387 I409 3558	0I20 4062 6326 77I0	56
57	9405 6847 8683 076I	0000 0000 0000 0000	0II0 8II7 I430 6476	57
58	0,95II 5838 387I 7I68	0,0202 0077 0665 8348	0,0I00 9563 7448 03I5	58
59	9607 5570 0058 8327	0000 0000 0000 0000	0090 97I8 8976 2772	59
60	9693 47I8 6305 5842	0I6I 50I7 93I6 I645	0080 8I58 I326 6994	60
6I	9769 0846 5774 I695	0000 0000 0000 0000	0070 3572 698I 0250	6I
62	9834 II45 4844 9964	0II9 5777 4050 8747	0059 69II 6927 2883	62
63	0,9888 4888 3I97 780I	0,0000 0000 0000 0000	0,0049 078I 74I0 643I	63
64	9932 2786 3II5 32I2	0076 6085 0756 4673	0038 4732 7993 77I2	64
65	9965 3067 7766 3027	0000 0000 0000 0000	0027 4947 5I50 3482	65
66	9987 I234 7I23 2276	0033 03II 3923 7937	00I6 I254 4764 3I47	66
67	9997 8624 0200 9853	0000 0000 0000 0000	0005 7580 0645 024I	67
Remainder terms		$G_{66} = 0{,}142 \times 10^{-17}$	$K_{102} = 0{,}213 \times 10^{-31}$	

$$n = 41$$

No.	Nodes Mantissa	Order	Weights of Gaussian quadratures Mantissa	Order	Weights of improved quadratures Mantissa	Order	No.
43	4137 7656 3203 7326	0	0000 0000 0000 0000	0	5775 1641 4571 5067	− 5	43
44	4277 6032 2355 7053	0	5763 0647 3151 2736	− 4	5763 2225 7726 2403	− 5	44
45	4437 0707 6632 7003	0	0000 0000 0000 0000	0	5742 2030 5150 0126	− 5	45
46	4575 6476 1431 1744	0	5712 5600 1347 5503	− 4	5712 4202 4370 4775	− 5	46
47	4733 5506 1454 1350	0	0000 0000 0000 0000	0	5654 4270 0745 2013	− 5	47
50	5070 4261 7112 5710	0	5607 6343 6464 0654	− 4	5607 7773 1542 6511	− 5	50
51	5224 1037 0742 6645	0	0000 0000 0000 0000	0	5534 4501 6736 2027	− 5	51
52	5356 2064 3332 1153	0	5452 7472 7372 1666	− 4	5452 5775 0414 2352	− 5	52
53	5506 5745 4303 1251	0	0000 0000 0000 0000	0	5363 0241 5263 0321	− 5	53
54	5635 3070 5267 0854	0	5264 7623 7013 1714	− 4	5265 1407 0134 7677	− 5	54
55	5762 0000 7336 0356	0	0000 0000 0000 0000	0	5160 7044 6271 3607	− 5	55
56	6104 3036 6450 4566	0	5046 7205 3173 1436	− 4	5046 5311 6441 3702	− 5	56
57	6224 2717 2663 4077	0	0000 0000 0000 0000	0	4727 1566 0621 2173	− 5	57
60	6341 6153 6460 0063	0	4602 0011 2647 1607	− 4	4602 2045 6141 5307	− 5	60
61	6454 3425 5361 3584	0	0000 0000 0000 0000	0	4447 3414 0611 4437	− 5	61
62	6564 1376 1305 5240	0	4307 5517 5570 1606	− 4	4307 3260 1243 1740	− 5	62
63	6670 6710 4073 7313	0	0000 0000 0000 0000	0	4143 0166 2750 6607	− 5	63
64	6772 2477 5606 4620	0	7763 2425 0313 7200	− 5	7763 7671 6651 5030	− 6	64
65	7070 3350 1753 6062	0	0000 0000 0000 0000	0	7427 4716 5812 5510	− 6	65
66	7163 0123 7733 5537	0	7063 2546 0052 0576	− 5	7062 4310 2730 3060	− 6	66
67	7251 7650 1110 2042	0	0000 0000 0000 0000	0	6507 0332 5632 4672	− 6	67
70	7335 1502 5057 1356	0	6123 3221 2310 2570	− 5	6124 3014 0457 7545	− 6	70
71	7414 4430 4171 3640	0	0000 0000 0000 0000	0	5530 6715 1625 4023	− 6	71
72	7467 7435 5200 5624	0	5127 6026 1272 1744	− 5	5126 4054 1257 5371	− 6	72
73	7537 2025 7163 1302	0	0000 0000 0000 0000	0	4520 6140 2215 7604	− 6	73
74	7602 3443 0772 3360	0	4104 6541 3771 4512	− 5	4106 4233 6776 0576	− 6	74
75	7641 3254 2572 7603	0	0000 0000 0000 0000	0	7150 5764 4224 1147	− 7	75
76	7674 0332 3121 1057	0	6076 5212 0755 0120	− 6	6071 4224 7750 6724	− 7	76
77	7722 2463 2340 2505	0	0000 0000 0000 0000	0	5015 0701 5060 0165	− 7	77
100	7744 2056 3130 7147	0	7660 3737 5653 1072	− 7	7702 1563 7220 2376	−10	100
101	7761 6242 3347 3614	0	0000 0000 0000 0000	0	5503 0211 6641 3313	−10	101
102	7772 5634 6120 0563	0	6607 4416 2161 2171	−10	6465 6006 0325 4277	−11	102
103	7777 0775 5512 3556	0	0000 0000 0000 0000	0	4557 0523 6037 0215	−12	103

No.	Nodes	Weights of Gaussian quadratures	Weights of improved quadratures	No.
I	0,0002 0I49 2I07 4670	0,0000 0000 0000 0000	0,0005 4283 2245 4523	I
2	00I2 I4I2 3I04 5790	003I I457 0277 9543	00I5 2067 7549 I8I3	2
3	0082 7I48 8679 245I	0000 0000 0000 0000	0025 9277 434I 5I76	3
4	0063 8609 I796 8453	0072 2508 I374 2975	0036 2822 0653 8080	4
5	0I05 I6I4 0708 6I98	0000 0000 0000 0000	0046 2940 4563 3754	5
6	0,0I56 4586 8733 3279	0,0II2 8I86 0992 7475	0,0056 3204 69I9 9668	6
7	02I7 8228 5333 4895	0000 0000 0000 0000	0066 3973 4673 3777	7
8	0289 I880 I297 4465	0I52 4569 03I9 223I	0076 2858 0794 60I6	8
9	0370 3023 6264 733I	0000 0000 0000 0000	0085 906I 5866 7090	9
I0	0460 95I6 II40 8378	0I90 8329 6898 I938	0095 376I I697 6506	I0
II	0,056I 0I70 3I67 64I8	0,0000 0000 0000 0000	0,0I04 7270 7396 4II7	II
I2	0670 3268 0832 7I78	0227 6280 576I 6766	0II3 8439 2002 3586	I2
I3	0788 6050 6638 2252	0000 0000 0000 0000	0I22 6662 0995 8486	I3
I4	09I5 5788 6049 5332	0262 5370 7286 339I	0I3I 2455 9566 0I54	I4
I5	I05I 02I9 2390 2354	0000 0000 0000 0000	0I39 598I 6780 0603	I5
I6	0,II94 6756 I685 0635	0,0295 2706 79I3 7622	0,0I47 6533 3805 2240	I6
I7	I346 2I37 I492 6205	0000 0000 0000 0000	0I55 8669 I96I 43I3	I7
I8	I505 3044 339I 8685	0325 5576 0777 0382	0I62 7644 7803 6I25	I8
I9	I67I 6404 8606 5679	0000 0000 0000 0000	0I69 8530 4854 9583	I9
20	I844 89I3 6459 7857	0358 I468 7907 I279	0I76 5849 II64 8265	20
2I	0,2024 6800 6I20 0459	0,0000 0000 0000 0000	0,0I82 9280 3480 6770	2I
22	22I0 6224 9665 I267	0377 8098 7330 0I60	0I88 8957 55I9 I488	22
23	2402 3476 6308 0667	0000 0000 0000 0000	0I94 490I 8230 I393	23
24	2599 4672 7404 8365	0399 3422 2I69 8859	0I99 6783 8999 2480	24
25	280I 56I2 378I 803I	0000 0000 0000 0000	0204 4382 7897 0029	25
26	0,3008 2036 II20 6770	0,04I7 5654 9849 9228	0,0208 7770 9530 86I9	26
27	32I8 9752 5223 4762	0000 0000 0000 0000	02I2 6944 24I3 0974	27
28	3433 4445 9330 2684	0432 3286 9873 5I79	02I6 I685 7282 6488	28
29	365I I599 9524 6I26	0000 0000 0000 0000	02I9 I862 8443 I269	29
30	387I 6665 4I9I 7753	0448 5094 89I7 8469	022I 75I8 I552 7398	30
3I	0,4094 5I25 2357 6006	0,0000 0000 0000 0000	0,0223 8635 56I5 6I7I	3I
32	43I9 2382 I370 4085	045I 0I52 2I85 3204	0225 5093 3463 2339	32
33	4545 374I 3779 5032	0000 0000 0000 0000	0226 6842 8300 4I88	33
34	4772 4508 9023 4487	0454 7837 0I65 I299	0227 39I2 8I73 I339	34
35	0,5000 0000 0000 0000	0,0000 0000 0000 0000	0,0227 628I 0925 0567	35

No.	Nodes		Weights of Gaussian quadratures		Weights of improved quadratures		No.
	Mantissa	Order	Mantissa	Order	Mantissa	Order	
I	6464 3640 I420 5I20	-I4	0000 0000 0000 0000	0	4344 6332 7563 475I	-I2	I
2	4762 I466 034I I202	-II	630I 672I 3702 I346	-I0	6I65 0570 5855 4I53	-II	2
3	6546 3I70 7057 3662	-I0	0000 0000 0000 0000	0	5286 56II 00I3 I243	-I0	3
4	6424 II53 5525 3220	- 7	73I4 0I40 0665 2576	- 7	7334 356I 03I7 7667	-I0	4
5	5304 5744 066I 6457	- 6	0000 0000 0000 0000	0	4573 II02 4630 7044	- 7	5
6	4002 5703 64I5 7I56	- 5	56I5 36I6 000I 7235	- 6	56I0 64I0 5245 3574	- 7	6
7	5447 0303 3I07 3232	- 5	0000 0000 0000 0000	0	663I I04I 5065 6365	- 7	7
I0	73I6 3437 I637 6I32	- 5	7634 44I7 25I5 65I3	- 6	7637 4454 4070 5754	- 7	I0
II	4572 6404 2670 2735	- 4	0000 0000 0000 0000	0	48I3 7647 48I4 6407	- 6	II
I2	57I4 7I07 4632 4363	- 4	4705 2223 0047 5663	- 5	4704 I644 45I3 0736	- 6	I2
I3	7I34 5346 2I32 6702	- 4	0000 0000 0000 0000	0	527I 2667 7422 4637	- 6	I3
I4	4224 4I56 0770 2220	- 3	5647 4420 2237 4243	- 5	5650 263I 6527 0066	- 6	I4
I5	5030 0636 0I72 4I07	- 3	0000 0000 0000 0000	0	62I7 4760 00II 6473	- 6	I5
I6	5670 I263 3443 0I5I	- 3	656I I003 6I43 0725	- 5	6560 4I44 4I6I 0563	- 6	I6
I7	6563 772I 3606 7203	- 3	0000 0000 0000 0000	0	7II3 3667 I035 I440	- 6	I7
20	75I2 5550 5457 3060	- 3	7436 I277 7022 5005	- 5	7436 5II4 3473 2636	- 6	20
2I	4235 5057 237I 6447	- 2	0000 0000 0000 0000	0	7750 6633 7563 0440	- 6	2I
22	4642 2247 0332 72I3	- 2	4I25 4460 3725 3450	- 4	4I25 3057 3005 730I	- 5	22
23	5262 64I5 3I65 2042	- 2	0000 0000 0000 0000	0	4262 2304 0672 5430	- 5	23
24	57I6 5270 2743 7524	- 2	44I2 3042 264I I256	- 4	44I2 42I2 203I 3283	- 5	24
25	6365 I705 706I 3I63	- 2	0000 0000 0000 0000	0	4585 58I2 0547 2004	- 5	25
26	7045 7044 3454 2203	- 2	4654 0074 4376 I585	- 4	4653 7II7 5056 0646	- 5	26
27	7540 0032 205I 7423	- 2	0000 0000 0000 0000	0	4765 I6I4 II7I 5662	- 5	27
30	4I2I 3674 6284 7737	- I	507I I02I I065 7654	- 4	507I I627 7362 350I	- 5	30
3I	4367 0237 4655 2575	- I	0000 0000 0000 0000	0	5I67 4720 4I62 006I	- 5	3I
32	4640 2440 2637 3753	- I	5260 4352 3777 6305	- 4	5260 3672 75I5 0I46	- 5	32
33	5II4 7700 4757 74I7	- I	0000 0000 0000 0000	0	5343 6500 7443 0I47	- 5	33
34	5374 5330 2402 6060	- I	542I 2363 I070 0364	- 4	542I 2725 7044 3744	- 5	34
35	5657 0I73 7646 4274	- I	0000 0000 0000 0000	0	5470 7262 0254 I725	- 5	35
36	6I43 5265 I273 277I	- I	5532 4527 I446 0447	- 4	5532 427I 7432 4400	- 5	36
37	6482 I630 I505 650I	- I	0000 0000 0000 0000	0	5566 I627 II75 63II	- 5	37
40	6722 2436 3744 6I52	- I	56I3 6I37 53II I202	- 4	56I3 6274 I560 035I	- 5	40
4I	72I3 4440 5737 5626	- I	0000 0000 0000 0000	0	5633 I443 6I47 336I	- 5	4I
42	7505 4567 706I 02I4	- I	5644 3607 02I0 67I0	- 4	5644 3550 3657 5457	- 5	42
43	4000 0000 0000 0000	0	0000 0000 0000 0000	0	5647 4423 037I 7354	- 5	43

No.	Nodes	Weights of Gaussian quadratures	Weights of improved quadratures	No.
36	0,5227 549I 0976 55I3	0,0454 7837 0I65 I299	0,0227 39I2 8I73 I339	36
37	5454 6258 6220 4968	0000 0000 0000 0000	0226 6842 8300 4I88	37
38	5680 76I7 8629 59I5	045I 0I52 2I85 3204	0225 5093 3463 2339	38
39	5905 4874 7642 3994	0000 0000 0000 0000	0223 8635 56I5 6I7I	39
40	0,6I28 3334 5808 2247	0,0443 5094 89I7 8469	0,022I 75I8 I552 7398	40
4I	6348 8400 0475 3874	0000 0000 0000 0000	02I9 I862 8443 I269	4I
42	6566 5554 0669 73I6	0432 3286 9873 5I79	02I6 I685 7282 6488	42
43	678I 0247 4776 5238	0000 0000 0000 0000	02I2 6944 24I3 0974	43
44	699I 7963 8879 8230	04I7 5654 9849 9228	0208 7770 9530 86I9	44
45	0,7I98 4887 62I8 I969	0,0000 0000 0000 0000	0,0204 4382 7897 0029	45
46	7400 5327 2595 I685	0399 3422 2I69 8859	0I99 6783 8999 2430	46
47	7597 6528 369I 9333	0000 0000 0000 0000	0I94 490I 8230 I398	47
48	7789 3775 0334 8733	0377 8098 7330 0I60	0I88 8957 55I9 I438	48
49	7975 8I99 3879 954I	0000 0000 0000 0000	0I82 9280 3480 6770	49
50	0,8I55 I086 3540 2643	0,0353 I468 7907 I279	0,0I76 5849 II64 8265	50
5I	8328 3595 I393 432I	0000 0000 0000 0000	0I69 8530 4854 9583	5I
52	8494 6955 6608 I3I5	0325 5576 0777 0382	0I62 7644 7803 6I25	52
53	8653 7862 8507 3795	0000 0000 0000 0000	0I55 3669 I96I 43I3	53
54	8805 3243 83I4 9365	0295 2706 79I3 7622	0I47 6533 3805 2240	54
55	0,8948 9780 7609 7646	0,0000 0000 0000 0000	0,0I39 598I 6780 0603	55
56	9084 42II 3950 4668	0262 5370 7286 339I	0I3I 2455 9566 0I54	56
57	92II 3949 336I 7748	0000 0000 0000 0000	0I22 6662 0995 8486	57
58	9329 678I 9I67 2822	0227 6280 576I 6766	0II3 8439 2002 3586	58
59	9438 9829 6832 3582	0000 0000 0000 0000	0I04 7270 7396 4II7	59
60	0,9539 0483 8859 I622	0,0I90 8329 6898 I988	0,0095 376I I697 6506	60
6I	9629 6976 3735 2669	0000 0000 0000 0000	0085 906I 5866 7090	6I
62	97I0 8II9 8702 5535	0I52 4569 03I9 223I	0076 2858 0794 60I6	62
63	9782 I77I 4666 5I05	0000 0000 0000 0000	0066 8973 4673 3777	63
64	9843 54I3 I266 672I	0II2 8I86 0992 7475	0056 3204 69I9 9668	64
65	0,9894 8885 929I 3802	0,0000 0000 0000 0000	0,0046 2940 4563 3754	65
66	9936 I390 8203 I547	0072 2508 I374 2975	0036 2822 0653 8080	66
67	9967 285I I820 7549	0000 0000 0000 0000	0025 9277 434I 5I76	67
68	9987 8587 6895 42I0	003I I457 0277 9543	00I5 2067 7549 I8I3	68
69	9997 9850 7892 5330	0000 0000 0000 0000	0005 4288 2245 4523	69
	Remainder terms	$G_{68} = 0{,}365 \times 10^{-18}$	$K_{104} = 0{,}252 \times 10^{-32}$	

No.	Nodes		Weights of Gaussian quadratures		Weights of improved quadratures		No.
	Mantissa	Order	Mantissa	Order	Mantissa	Order	
44	4I35 I504 0347 367I	0	5644 3607 02I0 67I0	− 4	5644 3550 3657 5457	− 5	44
45	4272 I557 5020 I064	0	0000 0000 0000 0000	0	5633 Í443 6I47 336I	− 5	45
46	4426 6560 60I5 47I2	0	56I3 6I37 53II I202	− 4	56I3 6274 I560 035I	− 5	46
47	4562 7063 7I35 0537	0	0000 0000 0000 0000	0	5566 I627 II75 63II	− 5	47
50	47I6 I245 3242 2403	0	5532 4527 I446 0447	− 4	5532 427I 7432 4400	− 5	50
5I	5050 3702 0054 564I	0	0000 0000 0000 0000	0	5470 7262 0254 I725	− 5	5I
52	520I 5223 6576 4747	0	542I 2363 I070 0364	− 4	542I 2725 7044 3744	− 5	52
53	533I 4037 54I0 0I70	0	0000 0000 0000 0000	0	5343 6500 7443 0I47	− 5	53
54	5457 6557 6460 20I2	0	5260 4352 3777 6305	− 4	5260 3672 75I5 0I46	− 5	54
55	5604 3660 I45I 250I	0	0000 0000 0000 0000	0	5I67 4720 4I62 006I	− 5	55
56	5727 204I 466I 4020	0	507I I02I I065 7654	− 4	507I I627 7362 350I	− 5	56
57	6047 777I 3365 4073	0	0000 0000 0000 0000	0	4765 I6I4 II7I 5662	− 5	57
60	6I66 4I66 7064 7337	0	4654 0074 4376 I535	− 4	4653 7II7 5056 0646	− 5	60
6I	6302 54I6 4I63 5I43	0	0000 0000 0000 0000	0	4535 58I2 0547 2004	− 5	6I
62	64I4 252I 7207 0052	0	44I2 3042 264I I256	− 4	44I2 42I2 203I 3233	− 5	62
63	6523 2274 5I42 5367	0	0000 0000 0000 0000	0	4262 2304 0672 5430	− 5	63
64	6627 3326 I7II 2I35	0	4I25 4460 3725 3450	− 4	4I25 3057 3005 730I	− 5	64
65	6730 4564 I30I 4266	0	0000 0000 0000 0000	0	7750 6633 7563 0440	− 6	65
66	7026 5222 7232 047I	0	7436 I277 7022 5005	− 5	7436 5II4 3473 2636	− 6	66
67	7I2I 4005 64I7 I057	0	0000 0000 0000 0000	0	7II3 3667 I035 I440	− 6	67
70	72I0 765I 4433 4762	0	656I I003 6I43 0725	− 5	6560 4I44 4I6I 0563	− 6	70
7I	7274 77I4 I760 5367	0	0000 0000 0000 0000	0	62I7 4760 00II 6473	− 6	7I
72	7355 3362 I700 7555	0	5647 4420 2237 4243	− 5	5650 263I 6527 0066	− 6	72
73	7432 I52I 4672 2443	0	0000 0000 0000 0000	0	527I 2667 7422 4637	− 6	73
74	7503 I433 4I46 2560	0	4705 2223 0047 5663	− 5	4704 I644 45I3 0736	− 6	74
75	7550 2457 5644 3642	0	0000 0000 0000 0000	0	43I3 7647 43I4 6407	− 6	75
76	76II 4307 0I43 0035	0	7634 44I7 25I5 65I3	− 6	7637 4454 4070 5754	− 7	76
77	7646 6I7I 7II5 6II3	0	0000 0000 0000 0000	0	668I I04I 5065 6365	− 7	77
I00	7677 724I 7027 44I4	0	56I5 36I6 000I 7235	− 6	56I0 64I0 5245 3574	− 7	I00
I0I	7724 7320 337I I0I3	0	0000 0000 0000 0000	0	4573 II02 4630 7044	− 7	I0I
I02	7745 6573 I2II 2522	0	73I4 0I40 0665 2576	− 7	7334 356I 03I7 7667	−I0	I02
I03	7762 463I 4I6I 64I0	0	0000 0000 0000 0000	0	5236 56II 00I3 I243	−I0	I03
I04	7773 0I56 3II7 4366	0	630I 672I 3702 I346	−I0	6I65 0570 5355 4I53	−II	I04
I05	7777 I3I3 4I37 6357	0	0000 0000 0000 0000	0	4344 6332 7563 475I	−I2	I05

No.	Nodes	Weights of Gaussian quadratures	Weights of improved quadratures	No.
I	0,000I 9085 07I9 706I	0,0000 0000 0000 0000	0,0005 I275 4555 8788	I
2	00II 467I 5450 I999	0029 4I7I 67I0 22I5	00I4 86I8 0007 2854	2
8	0080 8985 8480 5454	0000 0000 0000 0000	0024 4904 5445 I58I	8
4	0060 82II 7778 0742	0068 254I 4I74 I807	0084 2774 8609 892I	4
5	0099 84I7 I074 8295	0000 0000 0000 0000	0043 740I 7888 9485	5
6	0,0I47 8II9 I980 885I	0,0I06 6I48 9955 74I8	0058 2206 8880 40I8	6
7	0205 8065 I502 0785	0000 0000 0000 0000	0062 7606 98I5 8097	7
8	0278 2742 5896 0868	0I44 I468 0054 447I	0072 I807 48I2 6468	8
9	0849 98I2 8474 6470	0000 0000 0000 0000	008I 2488 5999 9249	9
I0	0485 7286 9820 84I2	0I80 5505 798I 78I7	0090 2882 5564 7794	I0
II	0,0580 4I84 7080 4758	0,0000 0000 0000 0000	0,0099 I28I 5865 9628	II
I2	0688 9048 7487 8888	02I5 542I II68 085I	0I07 8086 450I 4I04	I2
I8	0745 9822 7659 4542	0000 0000 0000 0000	0II6 2090 5447 7888	I8
I4	0866 2505 0458 8878	0248 8468 5200 6768	0I24 8969 4982 4895	I4
I5	0994 6689 8487 I47I	0000 0000 0000 0000	0I82 8986 49I9 6228	I5
I6	0,II80 9487 8856 5487	0,0280 2040 8I06 I85I	0,0I40 I242 9685 2402	I6
I7	I274 805I 2I66 7966	0000 0000 0000 0000	0I47 5865 6470 24I9	I7
I8	I425 9274 922I 6856	0809 8688 5988 040I	0I54 6649 2845 4468	I8
I9	I584 0479 0755 92I7	0000 0000 0000 0000	0I6I 5287 4887 4802	I9
20	I748 878I 7667 0548	0886 III4 2684 5485	0I68 0727 48I8 8975	20
2I	0,I920 07I4 4756 889I	0,0000 0000 0000 0000	0,0I74 2588 8I44 9088	2I
22	2097 2782 7625 II77	0860 2289 7886 2800	0I80 0966 0582 2I62	22
28	2280 I582 4I5I 8709	0000 0000 0000 0000	0I85 6I78 90I8 8747	28
24	2468 886I 8879 2557	088I 5I72 8577 72I0	0I90 7727 6969 2259	24
25	266I 5690 8269 2852	0000 0000 0000 0000	0I95 5265 7582 8888	25
26	0,2859 8I22 924I 0929	0,0899 8247 II2I I62I	0,0I99 899I 7480 4674	26
27	806I 2465 I986 0788	0000 0000 0000 0000	0208 9067 2879 2965	27
28	8266 9922 2784 5980	04I5 0029 6854 4288	0207 5I89 5570 5525	28
29	8476 I299 9264 4748	0000 0000 0000 0000	02I0 690I I448 7II9	29
80	8688 2852 9895 8520	0426 9882 6696 0496	02I8 4546 7422 2469	80
8I	0,8902 9087 0792 49I0	0,0000 0000 0000 0000	0,02I5 8252 8060 0558	8I
82	4II9 7446 94I7 0052	0485 5222 8498 59I8	02I7 7727 2708 4866	82
88	4888 8086 4698 29I7	0000 0000 0000 0000	02I9 2707 7462 2986	88
84	4558 I482 8862 I704	0440 7026 52I5 I877	0220 8899 I788 4680	84
85	4778 8479 60I9 76I8	0000 0000 0000 0000	022I 0004 8762 9495	85
86	0,5000 0000 0000 0000	0,0442 4889 7458 552I	0,022I 2288 2860 528I	86

n = 48

No.	Nodes Mantissa	Order	Weights of Gaussian quadratures Mantissa	Order	Weights of improved quadratures Mantissa	Order	No.
I	6I7I 434I 6654 I0I5	-I4	0000 0000 0000 0000	0	4I46 5I40 I364 7767	-I2	I
2	4544 6542 5547 660I	-II	60I4 472I 04I0 I364	-I0	5708 6206 6476 4026	-II	2
8	6247 7443 5I87 5432	-I0	0000 0000 0000 0000	0	50I0 0052 9576 8464	-I0	8
4	6I82 4422 2250 666I	- 7	6772 867I 2507 I4I5	- 7	70I2 I776 5470 7675	-I0	4
5	5054 I856 7602 I244	- 6	0000 0000 0000 0000	0	4865 I752 6855 7564	- 7	5
6	7442 6320 0288 2250	- 6	5852 6607 5277 0I24	- 6	5346 2264 0524 0686	- 7	6
7	52II 430I II03 8665	- 5	0000 0000 0000 0000	0	6882 8574 I658 5400	- 7	7
IO	6775 6704 0408 6742	- 5	7302 5527 I258 64I6	- 6	7305 5647 I066 I077	- 7	IO
II	4365 506I I676 5556	- 4	0000 0000 0000 0000	0	4I2I 7075 7285 0546	- 6	II
I2	5447 4567 027I 2I67	- 4	4476 4063 3260 I567	- 5	4475 3222 03I6 5678	- 6	I2
I3	6624 II50 3207 5457	- 4	0000 0000 0000 0000	0	5046 3508 45I4 I704	- 6	I3
I4	4035 I830 876I 46I0	- 3	54II II65 0022 5476	- 5	54I2 0040 I748 6667	- 6	I4
I5	46I4 2I25 4246 35I4	- 3	0000 0000 0000 0000	0	5746 2684 I544 2I86	- 6	I5
I6	5426 4I7I 8588 2782	- 3	6275 5867 5067 80I7	- 5	6274 7784 564I 02II	- 6	I6
I7	6273 24II 2556 4257	- 3	0000 0000 0000 0000	0	66I6 4758 6545 7648	- 6	I7
20	7I7I 7III 0I07 44II	- 3	7I30 54I6 0337 4360	- 5	7I8I 2I48 6025 2678	- 6	20
2I	4050 5I00 2650 4856	- 2	0000 0000 0000 0000	0	7433 4522 0044 7626	- 6	2I
22	4440 I725 3248 845I	- 2	7726 7477 2628 7I68	- 5	7726 3454 I72I 784I	- 6	22
28	5043 2885 4864 7075	- 2	0000 0000 0000 0000	0	4I05 I4I0 5I28 4I7I	- 5	28
24	546I 27I2 6I26 2456	- 2	4282 5726 8I45 4II2	- 4	4282 7550 7522 I554	- 5	24
25	6III 6605 2724 7840	- 2	0000 0000 0000 0000	0	4858 7660 0I47 5806	- 5	25
26	6554 I802 55I6 4820	- 2	4470 6070 5262 4522	- 4	4470 4876 6776 74I4	- 5	26
27	7227 6872 7854 7648	- 2	0000 0000 0000 0000	0	4600 78II 6024 7III	- 5	27
30	77I4 I502 7366 5I42	- 2	4704 2874 5206 7524	- 4	4704 3768 0588 7868	- 5	30
3I	4204 2667 7028 6786	- I	0000 0000 0000 0000	0	5002 6845 0678 8I26	- 5	3I
32	4446 2628 77I7 6678	- I	5074 2250 6740 737I	- 4	5074 0745 I472 I280	- 5	32
88	47I8 6I86 6480 7446	- I	0000 0000 0000 0000	0	5I60 5I26 7006 8242	- 5	88
84	5I64 2436 6657 5025	- I	5287 6067 07I6 4604	- 4	5287 7824 5244 6888	- 5	84
35	5437 5I24 5760 I802	- I	0000 0000 0000 0000	0	58II 4858 8I08 I527	- 5	35
36	57I5 8I60 0227 7445	- I	5855 7462 4752 5757	- 4	5855 6260 4876 II40	- 5	36
87	6I75 2064 8677 8I24	- I	0000 0000 0000 0000	0	54I4 6724 6552 48I0	- 5	87
40	6456 7I2I 2485 6660	- I	5446 I720 7630 I466	- 4	5446 8I00 2050 8588	- 5	40
4I	674I 7408 86I8 6447	- I	0000 0000 0000 0000	0	5472 0I52 046I 7050	- 5	4I
42	7226 0I76 7058 5072	- I	55I0 I405 5728 I565	- 4	55I0 024I I7I2 4226	- 5	42
48	75I2 6520 7I05 I056	- I	0000 0000 0000 0000	0	5520 545I 2582 4862	- 5	48
44	4000 0000 0000 0000	0	5528 4220 4440 8756	- 4	5528 586I 8502 7784	- 5	44

No.	Nodes	Weights of Gaussian quadratures	Weights of improved quadratures	No.
37	0,5221 1520 3980 2382	0,0000 0000 0000 0000	0,0221 0004 8762 9495	37
38	5441 8567 1637 8296	0440 7026 5215 1377	0220 3399 1733 4680	38
39	5661 6963 5306 7083	0000 0000 0000 0000	0219 2707 7462 2986	39
40	5880 2558 0582 9948	0435 5222 3498 5918	0217 7727 2708 4866	40
41	6097 0912 9207 5090	0000 0000 0000 0000	0215 8252 3060 0553	41
42	0,6311 7647 0604 6480	0,0426 9332 6696 0496	0,0213 4546 7422 2469	42
43	6523 8700 0785 5252	0000 0000 0000 0000	0210 6901 1448 7119	43
44	6733 0077 7215 4070	0415 0029 6864 4283	0207 5139 5570 5525	44
45	6938 7534 8013 9212	0000 0000 0000 0000	0203 9067 2379 2965	45
46	7140 6877 0758 9071	0399 8247 1121 1621	0199 8991 7430 4674	46
47	0,7338 4309 1730 7648	0,0000 0000 0000 0000	0,0195 5265 7582 3333	47
48	7531 6138 6620 7443	0381 5172 8577 7210	0190 7727 6969 2259	48
49	7719 8417 5848 1291	0000 0000 0000 0000	0185 6173 9018 3747	49
50	7902 7267 2374 8822	0360 2239 7386 2800	0180 0966 0532 2162	50
51	8079 9285 5243 6109	0000 0000 0000 0000	0174 2538 8144 9083	51
52	0,8251 1218 2332 9452	0,0336 1114 2634 5435	0,0168 0727 4813 8975	52
53	8415 9520 9244 0783	0000 0000 0000 0000	0161 5287 4837 4302	53
54	8574 0725 0778 3144	0309 3683 5983 0401	0154 6649 2845 4463	54
55	8725 1948 7833 2034	0000 0000 0000 0000	0147 5365 6470 2419	55
56	8869 0512 6143 4563	0280 2040 8106 1851	0140 1242 9635 2402	56
57	0,9005 3360 6562 8528	0,0000 0000 0000 0000	0,0132 3936 4919 6223	57
58	9133 7494 9546 1127	0248 8468 5200 6768	0124 3969 4932 4895	58
59	9254 0677 2340 5458	0000 0000 0000 0000	0116 2090 5447 7333	59
60	9366 0956 2512 6112	0215 5421 1163 0851	0107 8036 4501 4104	60
61	9469 5815 2919 5247	0000 0000 0000 0000	0099 1231 5365 9628	61
62	0,9564 2713 0679 6588	0,0180 5505 7981 7317	0,0090 2332 5564 7794	62
63	9650 0187 6525 3530	0000 0000 0000 0000	0081 2488 5999 9249	63
64	9726 7257 4103 9137	0144 1463 0054 4471	0072 1307 4312 6468	64
65	9794 1934 8497 9215	0000 0000 0000 0000	0062 7606 9315 8097	65
66	9852 1880 8019 6149	0106 6148 9955 7418	0053 2206 3380 4018	66
67	0,9900 6582 8925 6705	0,0000 0000 0000 0000	0043 7401 7383 9485	67
68	9939 6788 2221 9257	0068 2541 4174 1807	0034 2774 3609 3921	68
69	9969 1014 6519 4546	0000 0000 0000 0000	0024 4904 5445 1581	69
70	9988 5328 4549 8001	0029 4171 6710 2215	0014 3613 0007 2354	70
71	9998 0964 9280 2939	0000 0000 0000 0000	0005 1275 4555 3733	71
Remainder terms		$G_{70} = 0{,}938 \times 10^{-19}$	$K_{108} = 0{,}305 \times 10^{-33}$	

No.	Nodes Mantissa	Order	Weights of Gaussian quadratures Mantissa	Order	Weights of improved quadratures Mantissa	Order	No.
45	4I32 4527 4335 3350	0	0000 0000 0000 0000	0	5520 545I 2532 4362	− 5	45
46	4264 7700 4352 I342	0	55I0 I405 5723 I565	− 4	55I0 024I I7I2 4226	− 5	46
47	44I7 0I76 2072 0554	0	0000 0000 0000 0000	0	5472 0I52 046I 7050	− 5	47
50	4550 4327 256I 0447	0	5446 I720 7630 I466	− 4	5446 3I00 2050 3533	− 5	50
5I	470I 2745 6040 2325	0	0000 0000 0000 0000	0	54I4 6724 6552 48I0	− 5	5I
52	503I 2307 7664 0I55	0	5355 7462 4752 5757	− 4	5855 6260 4376 II40	− 5	52
53	5I60 I325 5007 7236	0	0000 0000 0000 0000	0	53II 4353 3I03 I527	− 5	53
54	5305 6560 4450 I365	0	5237 6067 07I6 4604	− 4	5237 7324 5244 6383	− 5	54
55	5482 0720 4563 4I54	0	0000 0000 0000 0000	0	5I60 5I26 7006 3242	− 5	55
56	5554 6466 0030 0442	0	5074 2250 6740 737I	− 4	5074 0745 I472 I230	− 5	56
57	5675 6444 0366 0420	0	0000 0000 0000 0000	0	5002 6345 0673 3I26	− 5	57
60	60I4 7457 2I02 2547	0	4704 2374 5206 7524	− 4	4704 3763 0533 7363	− 5	60
6I	6I32 030I 2I04 6027	0	0000 0000 0000 0000	0	4600 73II 6024 7III	− 5	6I
62	6244 75I7 2454 27I3	0	4470 6070 5262 4522	− 4	4470 4376 6776 74I4	− 5	62
63	6355 4236 52I2 6I07	0	0000 0000 0000 0000	0	4353 7660 0I47 5306	− 5	63
64	6463 52I5 2352 3264	0	4232 5726 3I45 4II2	− 4	4232 7550 7522 I554	− 5	64
65	6567 I3I0 4702 6I60	0	0000 0000 0000 0000	0	4I05 I4I0 5I23 4I7I	− 5	65
66	6667 74I2 5I27 I065	0	7726 7477 2623 7I63	− 5	7726 3454 I72I 734I	− 6	66
67	6765 6557 7225 6704	0	0000 0000 0000 0000	0	7433 4522 0044 7626	− 6	67
70	7060 6066 6767 0336	0	7I30 54I6 0337 4360	− 5	7I3I 2I43 6025 2673	− 6	70
7I	7I50 4536 6522 I352	0	0000 0000 0000 0000	0	66I6 4753 6545 7643	− 6	7I
72	7235 I360 6424 4504	0	6275 5367 5067 30I7	− 5	6274 7734 564I 02II	− 6	72
73	73I6 3565 2358 I426	0	0000 0000 0000 0000	0	5746 2634 I544 2I36	− 6	73
74	7374 2644 740I 63I6	0	54II II65 0022 5476	− 5	54I2 0040 I748 6667	− 6	74
75	7446 573I 3627 4II5	0	0000 0000 0000 0000	0	5046 3503 45I4 I704	− 6	75
76	75I5 4I50 4364 3270	0	4476 4063 3260 I567	− 5	4475 3222 03I6 5673	− 6	76
77	7560 5I34 7804 05II	0	0000 0000 0000 0000	0	4I2I 7075 7235 0546	− 6	77
I00	7620 042I 6767 7020	0	7302 5527 I253 64I6	− 6	7305 5647 I066 I077	− 7	I00
I0I	7653 547I 7555 7I02	0	0000 0000 0000 0000	0	6332 3574 I653 5400	− 7	I0I
I02	7703 35I4 5775 4455	0	5352 6607 5277 0I24	− 6	5346 2264 0524 0636	− 7	I02
I03	7727 2364 2I0I 7565	0	0000 0000 0000 0000	0	4365 I752 6355 7564	− 7	I03
I04	7747 2255 6666 5844	0	6772 367I 2507 I4I5	− 7	70I2 I776 5470 7675	−I0	I04
I05	7763 2600 6705 5004	0	0000 0000 0000 0000	0	50I0 0052 0576 3464	−I0	I05
I06	7773 233I 2352 230I	0	60I4 472I 04I0 I364	−I0	5703 6206 6476 4026	−II	I06
I07	7777 I606 3436 II23	0	0000 0000 0000 0000	0	4I46 5I40 I364 7767	−I2	I07

No.	Nodes	Weights of Gaussian quadratures	Weights of improved quadratures	No.
I	0,000I 800I 9242 5608	0,0000 0000 0000 0000	0,0004 8498 536I 5284	I
2	00I0 8476 8757 957I	0027 8285 9832 I225	00I3 587I 8439 4546	2
3	0029 23I5 I657 58I7	0000 0000 0000 0000	0023 I697 582I 6265	3
4	0057 0676 0548 8939	0064 5797 3642 0328	0032 4300 6060 46I0	4
5	0098 9886 0008 6788	0000 0000 0000 0000	004I 39II 4493 297I	5
6	0,0I89 86I5 4475 I5I0	0,0I00 9075 7648 8677	0,0050 3742 7730 6696	6
7	0I94 7592 5888 5I65	0000 0000 0000 0000	0059 4I38 022I I78I	7
8	0258 6850 7800 2462	0I36 498I 0749 2844	0068 2979 7I69 44I5	8
9	038I 2777 II88 0569	0000 0000 0000 0000	0076 9570 3522 0693	9
I0	04I2 5III 2742 I705	0I7I 0690 5385 I536	0085 4982 5744 42I5	I0
II	0,0502 2469 I404 9I56	0,0000 0000 0000 0000	0,0093 95I8 6769 7I62	II
I2	0600 3509 9554 80I4	0204 3787 546I 8224	0I02 2I63 I57I 285I	I2
I3	0706 5973 8885 2049	0000 0000 0000 0000	0II0 2386 5963 8459	I3
I4	0820 764I 6503 7624	0236 I754 I745 I330	0II8 0669 5025 4472	I4
I5	0942 673I 9996 73I9	0000 0000 0000 0000	0I25 7I75 3I3I 27I0	I5
I6	0,I072 II88 4933 8967	0,0266 2285 6988 8800	0,0I33 I28I 5900 7559	I6
I7	I208 8356 2923 7369	0000 0000 0000 0000	0I40 2599 0898 7455	I7
I8	I352 554I 4203 22I7	0294 3007 2I22 6624	0I47 I372 3386 347I	I8
I9	I508 0286 8759 5276	0000 0000 0000 0000	0I53 7680 942I 0258	I9
20	I659 9988 I707 2395	0320 I989 8677 5077	0I60 II0I 0767 7386	20
2I	0,I823 I424 4I07 97I4	0,0000 0000 0000 0000	0,0I66 I845 87I6 I625	2I
22	I992 I6I7 0982 0097	0343 7266 I9I7 8682	0I7I 8547 00I7 3324	22
23	2I66 75I9 5298 4467	0000 0000 0000 0000	0I77 2736 70I7 8333	23
24	2346 5985 7036 8774	0364 7094 2502 8265	0I82 36I6 6932 87I0	24
25	258I 3577 096I I240	0000 0000 0000 0000	0I87 098I 57I0 8794	25
26	0,2720 6802 7788 2899	0,0382 9920 5322 9353	0,0I9I 4904 6795 648I	26
27	29I4 2244 4723 4202	0000 0000 0000 0000	0I95 5892 322I 3704	27
28	3III 6872 6440 I554	0398 439I 4456 0358	0I99 2289 I499 0295	28
29	33I2 546I 6493 I662	0000 0000 0000 0000	0202 58I0 4568 8803	29
30	35I6 5750 2327 9859	04I0 9363 3352 I699	0205 4649 080I I6II	30
3I	0,3723 35I5 0033 9255	0,0000 0000 0000 0000	0,0208 0247 7899 589I	3I
32	3932 4955 384I 5672	0420 39I0 9489 83I0	02I0 I978 I456 4II6	32
33	4I43 6I57 390I 4235	0000 0000 0000 0000	02II 977I 0I78 848I	33
34	4856 3I94 8095 3076	0426 7334 2869 6693	02I3 3653 7738 0004	34
35	4570 2I57 4977 8043	0000 0000 0000 0000	02I4 36I2 0986 97I3	35
36	0,4784 9090 0763 I457	0,0429 9I63 7885 I974	0,02I4 9586 3I02 5258	36
37	5000 0000 0000 0000	0000 0000 0000 0000	02I5 I57I I9I2 4400	37

No.	Nodes Mantissa	Order	Weights of Gaussian quadratures Mantissa	Order	Weights of improved quadratures Mantissa	Order	No.
I	5714 1614 1158 0750	-14	0000 0000 0000 0000	0	7744 2642 1046 4635	-13	I
2	4342 7314 5382 2325	-11	5546 0243 7576 1665	-10	5441 3406 4141 5687	-11	2
3	5771 1130 6123 2557	-10	0000 0000 0000 0000	0	4575 4147 3022 3571	-10	3
4	5657 7706 7548 4231	- 7	6471 6550 6272 2232	- 7	6510 4234 7606 6310	-10	4
5	4687 6655 0555 4144	- 6	0000 0000 0000 0000	0	4172 0550 5420 4704	- 7	5
6	7122 3057 0105 1461	- 6	5125 1664 4187 7772	- 6	5121 0401 5337 4752	- 7	6
7	4770 5772 0422 2334	- 5	0000 0000 0000 0000	0	6052 7750 6761 5755	- 7	7
10	6475 7665 0122 7105	- 5	6772 0533 6437 6725	- 6	6774 6175 5731 1607	- 7	10
11	4173 0374 2722 0455	- 4	0000 0000 0000 0000	0	7702 6075 3623 3507	- 7	11
12	5217 3355 2046 5561	- 4	4302 1707 7124 3660	- 5	4301 2221 3725 6314	- 6	12
13	6333 4147 7411 1250	- 4	0000 0000 0000 0000	- 0	4636 7104 7651 5610	- 6	13
14	7586 3535 2463 1432	- 4	5166 6524 6552 4770	- 5	5167 4241 2475 7030	- 6	14
15	4413 3015 4801 0444	- 3	0000 0000 0000 0000	0	5511 6561 7451 1272	- 6	15
16	5201 3656 1124 0224	- 3	6027 4623 1320 4110	- 5	6027 0336 2010 2544	- 6	16
17	6020 7471 4073 6600	- 3	0000 0000 0000 0000	0	6337 4701 1065 3450	- 6	17
20	6671 0747 4704 2320	- 3	6641 3441 0470 1170	- 5	6641 6777 1645 2240	- 6	20
21	7571 0715 1260 6487	- 3	0000 0000 0000 0000	0	7134 6505 0321 0021	- 6	21
22	4250 0145 1447 0147	- 2	7421 3525 5222 2352	- 5	7421 0724 1374 5305	- 6	22
23	4636 4376 6164 2303	- 2	0000 0000 0000 0000	0	7676 7403 3122 0334	- 6	23
24	5237 5675 7845 5362	- 2	4062 3514 0644 1020	- 4	4062 4605 7477 1150	- 5	24
25	5653 0225 6402 3640	- 2	0000 0000 0000 0000	0	4201 4362 5667 7751	- 5	25
26	6277 7522 7277 1406	- 2	4314 5131 1320 2863	- 4	4314 4212 7541 5511	- 5	26
27	6736 0032 1552 2434	- 2	0000 0000 0000 0000	0	4423 4373 5453 7274	- 5	27
30	7404 5254 3352 0463	- 2	4526 1216 0515 6854	- 4	4526 2003 4451 4057	- 5	30
31	4031 5403 0067 1630	- 1	0000 0000 0000 0000	0	4624 2523 6457 2416	- 5	31
32	4264 6200 1063 5687	- 1	4715 7640 5123 1185	- 4	4715 7166 1571 2134	- 5	32
33	4523 2522 4655 0236	- 1	0000 0000 0000 0000	0	5002 7614 4575 0446	- 5	33
34	4765 0332 0630 3037	- 1	5063 1587 2610 4140	- 4	5063 2110 3744 4474	- 5	34
35	5231 5064 4017 4556	- 1	0000 0000 0000 0000	0	5136 4726 5607 7677	- 5	35
36	5500 6163 5774 4222	- 1	5205 0714 1522 2143	- 4	5205 0435 2023 5720	- 5	36
37	5752 1266 3621 4013	- 1	0000 0000 0000 0000	0	5246 4765 2202 2330	- 5	37
40	6225 4001 3285 2415	- 1	5303 0463 4151 3256	- 4	5303 0655 1731 2121	- 5	40
41	6502 3463 2024 0133	- 1	0000 0000 0000 0000	0	5332 3321 7757 5717	- 5	41
42	6760 5446 4505 1400	- 1	5354 5076 2011 2140	- 4	5354 4766 3626 6466	- 5	42
43	7237 7273 2703 0461	- 1	0000 0000 0000 0000	0	5371 5315 6457 1321	- 5	43
44	7517 6302 3106 6056	- 1	5401 3777 7200 7362	- 4	5401 4027 5014 4735	- 5	44
45	4000 0000 0000 0000	0	0000 0000 0000 0000	0	5404 0667 7007 1547	- 5	45

No.	Nodes	Weights of Gaussian quadratures	Weights of improved quadratures	No.
38	0,52I5 0909 9236 8543	0,0429 9I63 7835 I974	0,02I4 9586 3I02 5258	38
39	0,5429 7842 5022 I957	0,0000 0000 0000 0000	0,02I4 36I2 0986 97I3	39
40	5643 6805 I904 6924	0426 7834 2869 6693	02I3 3653 7738 0004	40
4I	5856 3842 6098 5765	0000 0000 0000 0000	02II 977I 0I78 848I	4I
42	6067 5044 6I58 4328	0420 39I0 9489 88I0	02I0 I978 I456 4II6	42
43	6276 6484 9966 0745	0000 0000 0000 0000	0208 0247 7899 589I	43
44	0,6483 4249 7672 0I4I	0,04I0 9363 3352 I699	0,0205 4649 080I I6II	44
45	6687 4538 3506 8338	0000 0000 0000 0000	0202 53I0 4568 8303	45
46	6888 3627 3559 8446	0398 439I 4456 0358	0I99 2239 I499 0295	46
47	7085 7755 5276 5798	0000 0000 0000 0000	0I95 5392 322I 3704	47
48	7279 3I97 22I6 7I0I	0382 9920 5322 9853	0I9I 4904 6795 643I	48
49	0,7468 6422 9038 8760	0,0000 0000 0000 0000	0,0I87 098I 57I0 8794	49
50	7653 40I4 2963 I226	0364 7094 2502 8265	0I82 36I6 6932 87I0	50
5I	7833 2480 470I 5533	0000 0000 0000 0000	0I77 2736 70I7 8333	5I
52	8007 8382 9067 9903	0343 7266 I9I7 8682	0I7I 8547 00I7 3324	52
53	8I76 8575 5892 0286	0000 0000 0000 0000	0I66 I345 87I6 I625	53
54	0,8340 006I 8292 7605	0,0320 I989 8677 5077	0,0I60 II0I 0767 7386	54
55	8496 97I3 I240 4724	0000 0000 0000 0000	0I58 7680 942I 0258	55
56	8647 4458 5796 7783	0294 3007 2I22 6624	0I47 I372 3386 347I	56
57	879I I643 7076 263I	0000 0000 0000 0000	0I40 2599 0898 7455	57
58	8927 88II 5066 I033	0266 2235 6988 8800	0I33 I28I 5900 7559	58
59	0,9057 3268 0003 268I	0,0000 0000 0000 0000	0,0I25 7I75 3I3I 27I0	59
60	9I79 2358 3496 2377	0236 I754 I745 I330	0II8 0669 5025 4472	60
6I	9293 4026 I6I4 795I	0000 0000 0000 0000	0II0 2386 5963 8459	6I
62	9399 6490 0445 I986	0204 3787 546I 8224	0I02 2I63 I57I 285I	62
63	9497 7530 8595 0844	0,0000 0000 0000 0000	0098 95I8 6769 7I62	63
64	0,9587 4888 7257 8295	0,0I7I 0690 5385 I536	0,0085 4982 5744 42I5	64
65	9668 7222 88I6 943I	0000 0000 0000 0000	0076 9570 3522 0693	65
66	974I 3649 2I99 7538	0I36 493I 0749 2844	0068 2979 7I69 44I5	66
67	9805 2407 4I6I 4835	0000 0000 0000 0000	0059 4I38 022I I78I	67
68	9860 I384 5524 8490	0I00 9075 7648 8677	0050 3742 7730 6696	68
69	0,9906 0II3 999I 32I2	0,0000 0000 0000 0000	0,004I 39II 4493 297I	69
70	9942 9323 945I I06I	0064 5797 3642 0328	0032 4300 6060 46I0	70
7I	9970 7684 8342 4683	0000 0000 0000 0000	0023 I697 582I 6265	7I
72	9989 I523 I242 0429	0027 8285 9832 I225	00I3 587I 8439 4546	72
73	9998 I998 0757 4397	0000 0000 0000 0000	0004 8498 536I 5284	73
	Remainder terms	$G_{72} = 0{,}24I \times I0^{-I9}$	$K_{110} = 0{,}36I \times I0^{-34}$	

No.	Nodes Mantissa	Order	Weights of Gaussian quadratures Mantissa	Order	Weights of improved quadratures Mantissa	Order	No.
46	4I30 0636 6334 4750	0	540I 3777 7200 7362	− 4	540I 4027 50I4 4735	− 5	46
47	4260 0242 2436 3547	0	0000 0000 0000 0000	0	537I 53I5 6457 I32I	− 5	47
50	4407 5I54 5535 3I77	0	5354 5076 20II 2I40	− 4	5354 4766 3626 6466	− 5	50
5I	4536 6I46 2765 7722	0	0000 0000 0000 0000	0	5332 332I 7757 57I3	− 5	5I
52	4665 I777 226I 257I	0	5303 0463 4I5I 3256	− 4	5303 0655 I73I 2I2I	− 5	52
53	50I2 7244 6067 I772	0	0000 0000 0000 0000	0	5246 4765 2202 2330	− 5	53
54	5I37 4706 I00I 5666	0	5205 07I4 I522 2I43	− 4	5205 0435 2023 5720	− 5	54
55	5263 I345 5770 I5I0	0	0000 0000 0000 0000	0	5I36 4726 5607 7677	− 5	55
56	5405 3622 7463 6360	0	5063 I537 26I0 4I40	− 4	5063 2II0 3744 4474	− 5	56
57	5526 2526 545I 3660	0	0000 0000 0000 0000	0	5002 76I4 4575 0446	− 5	57
60	5645 4677 7346 I060	0	47I5 7640 5I23 II85	− 4	47I5 7I66 I57I 2I34	− 5	60
6I	5763 II76 3744 3063	0	0000 0000 0000 0000	0	4624 2523 6457 24I6	− 5	6I
62	6076 6524 7I05 3663	0	4526 I2I6 05I5 6354	− 4	4526 2003 445I 4057	− 5	62
63	62I0 377I 3445 3270	0	0000 0000 0000 0000	0	4423 4373 5453 7274	− 5	63
64	6320 0053 2I20 I476	0	43I4 5I3I I320 2363	− 4	43I4 42I2 754I 55II	− 5	64
65	6425 I732 4277 3027	0	0000 0000 0000 0000	0	420I 4362 5667 775I	− 5	65
66	6530 0420 4I06 4503	0	4062 35I4 0644 I020	− 4	4062 4605 7477 II50	− 5	66
67	6630 2700 2342 73I7	0	0000 0000 0000 0000	0	7676 7403 3I22 0384	− 6	67
70	6725 7746 5466 I746	0	742I 3525 5222 2352	− 5	742I 0724 I374 5305	− 6	70
7I	7020 6706 265I 7I34	0	0000 0000 0000 0000	0	7I34 6505 032I 002I	− 6	7I
72	7II0 6703 0307 3545	0	664I 344I 0470 II70	− 5	664I 6777 I645 2240	−6	72
73	7I75 7030 6370 4I20	0	0000 0000 0000 0000	0	6337 470I I065 3450	− 6	73
74	7257 64I2 I665 3755	0	6027 4623 I320 4II0	− 5	6027 0336 20I0 2544	− 6	74
75	7336 4476 2347 6733	0	0000 0000 0000 0000	0	55II 656I 745I I272	− 6	75
76	74I2 06I2 I254 63I6	0	5I66 6524 6552 4770	− 5	5I67 424I 2475 7030	− 6	76
77	7462 2I7I 40I7 3325	0	0000 0000 0000 0000	0	4636 7I04 765I 56I0	− 6	77
I00	7527 022I I275 45I0	0	4302 I707 7I24 3660	− 5	430I 222I 3725 63I4	− 6	I00
I0I	7570 2360 I642 6755	0	0000 0000 0000 0000	0	7702 6075 3623 3507	− 7	I0I
I02	7626 0402 2575 32I5	0	6772 0533 6437 6725	− 6	6774 6I75 573I I607	− 7	I02
I03	7660 I640 I367 333I	0	0000 0000 0000 0000	0	6052 7750 676I 5755	− 7	I03
I04	7706 5547 2076 7263	0	5I25 I664 4I37 7772	− 6	5I2I 040I 5337 4752	− 7	I04
I05	778I 40II 2272 2236	0	0000 0000 0000 0000	0	4I72 0550 5420 4704	− 7	I05
I06	7750 5000 344I I6I6	0	647I 6550 6272 2282	− 7	65I0 4234 7606 63I0	−I0	I06
I07	7764 0I55 5I63 53I2	0	0000 0000 0000 0000	0	4575 4I47 3022 357I	−I0	I07
II0	7773 4350 4632 4455	0	5546 0243 7576 I665	−I0	544I 3406 4I4I 5637	−II	II0
III	7777 2063 6I63 6624	0	0000 0000 0000 0000	0	7744 2642 I046 4635	−I3	III

No.	Nodes	Weights of Gaussian quadratures	Weights of improved quadratures	No.
I	0,000I 7058 6620 I836	0,0000 0000 0000 0000	0,0004 5952 3283 9I07	I
2	00I0 2770 876I 0432	0026 3652 8639 7490	00I2 87I5 0573 9498	2
3	0027 6938 2855 63I8	0000 0000 0000 0000	002I 9526 7I20 7384	3
4	0054 070I 8392 8404	006I I939 0050 I538	0030 73I5 89I3 I7I4	4
5	0089 0590 537I 7006	0000 0000 0000 0000	0039 2263 4660 77I2	5
6	0,0I32 5348 497I 757I	0,0095 6452 2244 5420	0,0047 7449 4954 5600	6
7	0I84 573I 4966 3336	0000 0000 0000 0000	0056 3263 0I04 4462	7
8	0245 I382 8368 9526	0I29 430I 8495 2795	0064 7665 4360 4966	8
9	03I4 0328 9884 9224	0000 0000 0000 0000	0072 9942 3962 9740	9
I0	039I 0928 I293 768I	0I62 308I 9923 7607	008I II66 I450 8794	I0
II	0,0476 2450 9II2 0997	0,0000 0000 0000 0000	0,0089 I699 3I59 2763	II
I2	0569 375I 8922 2570	0I94 0480 I250 9673	0097 0530 0526 5693	I2
I3	0670 2737 5740 6I86	0000 0000 0000 0000	0I04 7088 2020 0726	I3
I4	0778 7850 6329 7220	0224 4268 233I 2I86	0II2 I899 I897 0I62	I4
I5	0894 6047 79II 6236	0000 0000 0000 0000	0II9 5206 I978 5860	I5
I6	0,I0I7 7039 9745 0489	0,0253 23I4 8827 4I23	0,0I26 6854 3488 I089	I6
I7	II47 7874 65I3 88I4	0000 0000 0000 0000	0I33 4890 8935 8344	I7
I8	I284 6058 3009 0I74	0280 2599 3999 I375	0I40 II29 9839 0946	I8
I9	I427 9437 3737 I253	0000 0000 0000 0000	0I46 5247 9592 II83	I9
20	I577 5684 5434 5203	0805 3225 826I 6I30	0I52 6762 4765 9960	20
2I	0,I738 I96I 6474 6980	0,0000 0000 0000 0000	0,0I58 580I 274I 3629	2I
22	I894 5369 5795 5378	0328 2436 I436 3756	0I64 I083 6496 3690	22
23	206I 3266 2827 8665	0000 0000 0000 0000	0I69 4254 5I0I 4456	23
24	2238 2880 4069 209I	0348 8622 5777 8502	0I74 4434 I045 849I	24
25	24I0 I025 5732 3642	0000 0000 0000 0000	0I79 I808 8250 2203	25
26	0,259I 4456 I098 8972	0,0367 0338 8624 244I	0,0I83 5055 5022 7487	26
27	2777 0I45 824I 6838	0000 0000 0000 0000	0I87 5809 8892 2746	27
28	2966 4974 5340 8369	0382 68I0 3785 2646	0I9I 3262 2850 I568	28
29	3I59 5473 9939 8467	0000 0000 0000 0000	0I94 7I44 7I5I 346I	29
30	3355 8I28 5058 I465	0395 5443 09I8 7647	0I97 76I9 5833 728I	30
3I	0,3554 9630 457I 5I06	0,0000 0000 0000 0000	0,0200 4828 3525 948I	3I
32	3756 66I0 3604 3I7I	0405 683I 2254 2325	0202 85I3 7985 72I5	32
33	3960 5400 6466 9255	0000 0000 0000 0000	0204 8444 7330 8373	33
34	4I66 2303 4880 0740	04I2 9763 6II8 2I86	0206 4786 I765 769I	34
35	4373 3834 7632 2I98	0000 0000 0000 0000	0207 7696 0436 2I55	35
36	0,458I 6479 5522 6I50	0,04I7 3728 68I2 93I4	0,0208 6958 4679 4228	36
37	4790 6463 0282 I358	0000 0000 0000 0000	0209 237I 8965 3079	37
38	5000 0000 0000 0000	04I8 84I8 0496 5695	0209 4II5 3794 4I24	38

No.	Nodes Mantissa	Order	Weights of Gaussian quadratures Mantissa	Order	Weights of improved quadratures Mantissa	Order	No.
I	5455 7577 2157 2412	-I4	0000 0000 0000 0000	0	7416 6052 0565 4027	-I3	I
2	4153 2057 0227 3756	-II	5314 4634 2205 7121	-I0	5213 2633 I730 I507	-II	2
3	5527 7210 6270 3733	-I0	0000 0000 0000 0000	0	4375 7I70 3725 I602	-I0	3
4	5422 6533 4524 I553	- 7	6210 2452 0430 2472	- 7	6226 3415 0I60 442I	-I0	4
5	4436 5023 0500 I656	- 6	0000 0000 0000 0000	0	40I0 456I 6250 2I74	- 7	5
6	6622 2445 I0I4 0027	- 6	4713 2203 4367 7700	- 6	4707 I535 6572 2764	- 7	6
7	4563 I7I3 4II6 0I00	- 5	0000 0000 0000 0000	0	56II 0755 0266 3I37	- 7	7
I0	6215 047I 3III 7376	- 5	6500 7364 2230 2430	- 6	6503 5035 2523 4737	- 7	I0
II	40I2 0274 34I2 7II0	- 4	0000 0000 0000 0000	0	7363 000I 4632 6647	- 7	II
I2	5003 04I5 6040 3536	- 4	4II7 3I77 0605 0I54	- 5	4II6 3306 I227 2277	- 6	I2
I3	606I 0752 6437 70I2	- 4	0000 0000 0000 0000	0	444I 4224 3775 5430	- 6	I3
I4	7223 3520 6657 57I5	- 4	4757 332I 2604 I735	- 5	4760 I373 0560 060I	- 6	I4
I5	4224 2646 0537 0354	- 3	0000 0000 0000 0000	0	5270 70I7 7530 4375	- 6	I5
I6	4767 6045 0564 4675	- 3	5575 4667 25I3 24I4	- 5	5574 7734 2223 60I4	- 6	I6
I7	5563 34I6 0565 0728	- 3	0000 0000 0000 0000	0	6075 I224 6505 3263	- 6	I7
20	6406 6377 65I5 4276	- 3	6367 II76 0027 22I3	- 5	6367 5300 22I0 0565	- 6	20
2I	726I 0436 4506 62I0	- 3	0000 0000 0000 0000	0	6653 254I 6502 0573	- 6	2I
22	4070 5453 6266 246I	- 2	7I3I 3304 7524 0406	- 5	7I30 7646 5I25 27I2	- 6	22
23	4443 4260 I520 I730	- 2	0000 0000 0000 0000	0	740I 0364 I235 I067	- 6	23
24	5030 5402 5367 I020	- 2	764I 73II 255I 6350	- 5	7642 24I7 2072 7360	- 6	24
25	5427 5262 4440 7I55	- 2	0000 0000 0000 0000	0	4035 7055 32I0 6433	- 5	25
26	6040 0046 2343 0522	- 2	4I47 I326 334I 4737	- 4	4I47 0004 6032 4472	- 5	26
27	646I 2I60 6536 7254	- 2	0000 0000 0000 0000	0	4254 5427 5076 2424	- 5	27
30	7II3 0II6 2556 6635	- 2	4356 2333 73I3 756I	- 4	4356 3557 2225 4643	- 5	30
3I	7554 5544 3627 600I	- 2	0000 0000 0000 0000	0	4453 7I70 0I50 7726	- 5	3I
32	4II2 7230 4333 246I	- I	4545 3II2 67I3 I064	- 4	4545 I747 I537 237I	- 5	32
33	4342 7342 5446 3052	- I	0000 0000 0000 0000	0	4632 5225 50I2 0725	- 5	33
34	4576 II7I 5523 3024	- I	47I3 4705 5546 5454	- 4	47I3 6004 5333 I464	- 5	34
35	5034 232I 654I 6633	- I	0000 0000 0000 0000	0	4770 I225 456I 0375	- 5	35
36	5275 05I7 32I7 I343	- I	5040 I723 5465 I750	- 4	5040 0660 2436 5267	- 5	36
37	5540 I634 4477 36I4	- I	0000 0000 0000 0000	0	5I03 6II4 2053 3534	- 5	37
40	6005 35I6 5646 2200	- I	5I42 5365 3I42 6I53	- 4	5I42 6404 4I74 5354	- 5	40
4I	6254 3627 I647 I5I7	- I	0000 0000 0000 0000	0	5I74 7377 727I 5I02	- 5	4I
42	6524 7635 2262 I757	- I	5222 3665 6207 5II3	- 4	5222 2664 30I0 2626	- 5	42
43	6776 58I7 66I3 5757	- I	0000 0000 0000 0000	0	5243 2I6I 5452 7277	- 5	43
44	725I 2223 3636 2565	- I	5257 2267 4742 5075	- 4	5257 3260 7533 6062	- 5	44
45	7524 3765 4367 3326	- I	0000 0000 0000 0000	0	5266 4070 0525 26I3	- 5	45
46	4000 0000 0000 0000	0	5270 7277 0563 66I4	- 4	5270 63I0 2540 00I7	- 5	46

No.	Nodes	Weights of Gaussian quadratures	Weights of improved quadratures	No.
39	0,5209 3536 9717 8642	0,0000 0000 0000 0000	0,0209 2371 8965 3079	39
40	5418 3520 4477 3850	0417 3728 6812 9314	0208 6958 4679 4228	40
41	0,5626 6165 2367 7802	0,0000 0000 0000 0000	0,0207 7696 0436 2155	41
42	5833 7696 5119 9260	0412 9763 6118 2186	0206 4786 1765 7691	42
43	6039 4599 3533 0745	0000 0000 0000 0000	0204 8444 7330 8373	43
44	6243 3889 6395 6829	0405 6831 2254 2825	0202 8513 7985 7215	44
45	6445 0869 5428 4894	0000 0000 0000 0000	0200 4828 3525 9481	45
46	0,6644 1871 4941 8585	0,0395 5443 0918 7647	0,0197 7619 5883 7281	46
47	6840 4526 0060 1533	0000 0000 0000 0000	0194 7144 7151 3461	47
48	7033 5025 4659 1631	0382 6810 3785 2646	0191 3262 2850 1568	48
49	7222 9854 1758 3162	0000 0000 0000 0000	0187 5809 8832 2746	49
50	7408 5543 8901 6028	0367 0338 8624 2441	0183 5055 5022 7487	50
51	0,7589 8974 4267 6358	0,0000 0000 0000 0000	0,0179 1308 8250 2203	51
52	7766 7119 5930 7909	0348 8622 5777 8502	0174 4434 1045 8491	52
53	7938 6733 7172 1334	0000 0000 0000 0000	0169 4254 5101 4456	53
54	8105 4630 4204 4622	0328 2436 1436 3756	0164 1083 6496 3690	54
55	8266 8038 3525 3020	0000 0000 0000 0000	0158 5301 2741 3629	55
56	0,8422 4315 4565 4797	0,0305 3225 8261 6130	0,0152 6762 4765 9960	56
57	8572 0562 6262 8747	0000 0000 0000 0000	0146 5247 9592 1183	57
58	8715 3941 6990 9826	0280 2599 3999 1375	0140 1129 9839 0946	58
59	8852 2125 3486 1186	0000 0000 0000 0000	0133 4890 8935 3344	59
60	8982 2960 0254 9511	0253 2314 8827 4123	0126 6354 3488 1089	60
61	0,9105 3952 2088 3764	0,0000 0000 0000 0000	0,0119 5206 1978 5860	61
62	9221 2649 3670 2780	0224 4268 2331 2186	0112 1899 1397 0162	62
63	9329 7262 4259 3864	0000 0000 0000 0000	0104 7088 2020 0726	63
64	9430 6248 1077 7431	0194 0480 1250 9673	0097 0530 0526 5693	64
65	9523 7549 0887 9003	0000 0000 0000 0000	0089 1699 3159 2763	65
66	0,9608 9071 8706 2319	0,0162 3081 9923 7607	0,0081 1166 1450 8794	66
67	9685 9671 0115 0776	0000 0000 0000 0000	0072 9942 3962 9740	67
68	9754 8617 1631 0474	0129 4301 8495 2795	0064 7665 4360 4966	68
69	9815 4268 5033 6664	0000 0000 0000 0000	0056 3263 0104 4462	69
70	9867 4651 5028 2429	0095 6452 2244 5420	0047 7449 4954 5600	70
71	0,9910 9409 4628 2994	0,0000 0000 0000 0000	0,0039 2263 4660 7712	71
72	9945 9298 1607 1596	0061 1939 0050 1538	0030 7315 8913 1714	72
73	9972 3061 7144 3682	0000 0000 0000 0000	0021 9526 7120 7384	73
74	9989 7229 1238 9568	0026 3652 8639 7490	0012 8715 0573 9499	74
75	9998 2941 3379 8164	0000 0000 0000 0000	0004 5952 3233 9107	75
	Remainder terms	$G_{74} = 0{,}620 \times 10^{-20}$	$K_{114} = 0{,}438 \times 10^{-35}$	

$$n = 45$$

No.	Nodes Mantissa	Order	Weights of Gaussian quadratures Mantissa	Order	Weights of improved quadratures Mantissa	Order	No.
47	4I25 6005 I604 2224	0	0000 0000 0000 0000	0	5266 4070 0525 26I3	− 5	47
50	4253 2666 2060 6505	0	5257 2267 4742 5075	− 4	5257 3260 7533 6062	− 5	50
5I	4400 5230 0472 I0I0	0	0000 0000 0000 0000	0	5243 2I6I 5452 7277	− 5	5I
52	4525 406I 2646 70I0	0	5222 3665 6207 5II3	− 4	5222 2664 30I0 2626	− 5	52
53	465I 6064 3054 3I30	0	0000 0000 0000 0000	− 0	5I74 7377 727I 5I02	− 5	53
54	4775 2I30 5054 6677	0	5I42 5365 3I42 6I53	− 4	5I42 6404 4I74 5854	− 5	54
55	5II7 706I 5540 207I	0	0000 0000 0000 0000	0	5I03 6II4 2053 3534	− 5	55
56	524I 3530 2270 32I6	0	5040 I723 5465 I750	− 4	5040 0660 2436 5267	− 5	56
57	586I 6627 05I7 0462	0	0000 0000 0000 0000	0	4770 I225 456I 0375	− 5	57
60	5500 7303 II26 2365	0	47I3 4705 5546 5454	− 4	47I3 6004 5333 I464	− 5	60
6I	56I6 42I6 5I54 6852	0	0000 0000 0000 0000	0	4632 5225 50I2 0725	− 5	6I
62	5782 4263 5622 2547	0	4545 3II2 67I3 I064	− 4	4545 I747 I537 237I	− 5	62
63	6044 6446 7032 0377	0	0000 0000 0000 0000	0	4453 7I70 0I50 7726	− 5	63
64	6I55 I754 3244 2230	0	4356 2333 73I3 756I	− 4	4356 3557 2225 4643	− 5	64
65	6263 5343 6250 2I24	0	0000 0000 0000 0000	0	4254 5427 5076 2424	− 5	65
66	6367 7766 3307 I653	0	4I47 I326 334I 4737	− 4	4I47 0004 6032 4472	− 5	66
67	6472 0523 2667 6I44	0	0000 0000 0000 0000	0	4035 7055 32I0 6433	− 5	67
70	657I 6477 2502 I573	0	764I 73II 255I 6350	− 5	7642 24I7 2072 7360	− 6	70
7I	6667 0723 7453 74II	0	0000 0000 0000 0000	0	740I 0364 I235 I067	− 6	7I
72	676I 6465 0322 3263	0	7I3I 3304 7524 0406	− 5	7I30 7646 5I25 27I2	− 6	72
73	705I 6734 I327 II56	0	0000 0000 0000 0000	0	6653 254I 6502 0573	− 6	73
74	7I37 II40 0I26 2350	0	6367 II76 0027 22I3	− 5	6367 5300 22I0 0565	− 6	74
75	722I 4436 I72I 2705	0	0000 0000 0000 0000	0	6075 I224 6505 3263	− 6	75
76	730I 0I73 272I 33I0	0	5575 4667 25I3 24I4	− 5	5574 7734 2223 60I4	− 6	76
77	7355 35I3 I724 0742	0	0000 0000 0000 0000	0	5270 70I7 7530 4375	− 6	77
I00	7426 62I2 7445 0I03	0	4757 332I 2604 I735	− 5	4760 I373 0560 060I	− 6	I00
I0I	7474 734I 2456 0037	0	0000 0000 0000 0000	0	444I 4224 3775 5430	− 6	I0I
I02	7537 6857 I075 76I2	0	4II7 3I77 0605 0I54	− 5	4II6 3306 I227 2277	− 6	I02
I03	7577 2764 I6I7 2433	0	0000 0000 0000 0000	0	7863 000I 4632 6647	− 7	I03
I04	7633 4566 I5I5 54I0	0	6500 7364 2230 2430	− 6	6503 5035 2523 4737	− 7	I04
I05	7664 3I4I 5075 4375	0	0000 0000 0000 0000	0	56II 0755 0266 3I37	− 7	I05
I06	77II 5553 3267 6377	0	47I3 2203 4367 7700	− 6	4707 I535 6572 2764	− 7	I06
I07	7733 4I27 5472 776I	0	0000 0000 0000 0000	0	40I0 456I 6250 2I74	− 7	I07
II0	775I 6645 22I5 257I	0	62I0 2452 0430 2472	− 7	6226 34I5 0I60 442I	−I0	II0
III	7764 520I 3563 2I70	0	0000 0000 0000 0000	0	4375 7I70 3725 I602	−I0	III
II2	7773 6245 7207 5504	0	53I4 4634 2205 7I2I	−I0	52I3 2633 I730 I507	−II	II2
II3	7777 2322 0200 5620	0	0000 0000 0000 0000	0	74I6 6052 0565 4027	−I3	II3

n = 38

No.	Nodes	Weights of Gaussian quadratures	Weights of improved quadratures	No.
I	0,000I 6I80 4408 56I5	0,0000 0000 0000 0000	0,0004 859I 50I9 7892	I
2	0009 7503 4732 I562	0025 OI44 0374 8I97	00I2 2I3I 4635 7948	2
3	0026 276I 0377 9304	0000 0000 0000 0000	0020 8292 7980 8920	3
4	005I 3027 2866 8072	0058 0672 2858 2343	0029 I596 9739 9869	4
5	0084 5046 3502 6058	0000 0000 0000 0000	0037 2264 3254 I033	5
6	0,0I25 7683 5704 9232	0,0090 7828 8854 8066	0,0045 3I98 7020 2685	6
7	0I75 I67I 8I24 37I8	0000 0000 0000 0000	0053 4726 3867 9452	7
8	0232 6683 4533 2852	0I22 8986 9869 II62	006I 4956 9680 7944	8
9	0298 0925 5092 8659	0000 0000 0000 0000	0069 8274 7447 5853	9
I0	037I 2933 3975 7078	0I54 I975 0272 5875	0077 0659 9729 4820	I0
II	0,0452 2047 8962 5288	0,0000 0000 0000 0000	0,0084 740I 2280 8I76	II
I2	0540 72I3 0497 6839	0I84 4704 0797 OI24	0092 2595 9237 7702	I2
I3	0636 6555 2I8I 7I4I	0000 0000 0000 0000	0099 5776 7I30 3I29	I3
I4	0739 8248 9033 8I89	02I3 5I57 9252 3372	0I06 7390 4070 7855	I4
I5	0850 0878 2600 30I3	0000 0000 0000 0000	0II3 7599 3395 5656	I5
I6	0,0967 279I 6I97 34I6	0,024I I403 0980 8793	0,0I20 5850 9I95 85I2	I6
I7	I09I I804 2049 I660	0000 0000 0000 0000	0I27 I800 2486 9775	I7
I8	I22I 5704 8I23 0I47	0267 I600 9955 I662	0I33 5680 0698 5453	I8
I9	I358 2509 6933 3687	0000 0000 0000 0000	0I39 7575 3703 8706	I9
20	I50I 0065 98I0 4078	029I 40I9 9573 4986	0I45 7I08 086I 9445	20
2I	0,I649 5846 9008 2832	0,0000 0000 0000 0000	0,0I5I 402I 0028 6476	2I
22	I803 7279 2085 I59I	03I3 7046 6696 0665	0I56 8442 957I 6605	22
23	I963 I920 4642 8048	0000 0000 0000 0000	0I62 04I4 0I04 6360	23
24	2I27 7I98 9476 0965	0333 9I96 8989 5702	0I66 9664 2963 I739	24
25	2297 0275 2276 03I8	0000 0000 0000 0000	0I7I 6004 750I II32	25
26	0,2470 8264 I086 0844	0,035I 9I25 3533 4495	0,0I75 9509 03I2 696I	26
27	2648 8355 752I I245	0000 0000 0000 0000	0I80 0I9I 4535 80I0	27
28	2830 764I 5283 8II8	0367 5634 6292 37I7	0I83 7860 4463 0093	28
29	30I6 3025 8I87 8543	0000 0000 0000 0000	0I87 2384 7087 6I48	29
30	3205 I377 9760 2825	0380 7683 I774 2232	0I90 3807 72I8 0068	30
3I	0,3396 9609 4222 2946	0,0000 0000 0000 0000	0,0I93 2I29 4268 I47I	3I
32	359I 4559 5I04 9I74	039I 4892 2329 I055	0I95 722I 6238 0672	32
33	3788 2948 32I5 7245	0000 0000 0000 0000	0I97 9005 4527 6832	33
34	3987 I477 3053 94I6	0399 505I 662I 7639	0I99 7508 0367 I860	34
35	4I87 6868 9697 2569	0000 0000 0000 0000	020I 272I 389I 5306	35
36	0,4389 5798 733I 0663	0,0404 9I24 6885 2986	0,0202 4572 8560 645I	36
37	4592 4883 79I4 4478	0000 0000 0000 0000	0203 3038 6945 9389	37
38	4796 0742 6047 7I09	0407 625I 4640 I929	0203 8I22 2549 6278	38
39	0,5000 0000 0000 0000	0,0000 0000 0000 0000	0,0203 9825 4853 6743	39

No.	Nodes Mantissa	Order	Weights of Gaussian quadratures Mantissa	Order	Weights of improved quadratures Mantissa	Order	No.
I	5232 50I2 2243 4I07	-I4	0000 0000 0000 0000	0	7II0 5606 0I75 40II	-I3	I
2	777I 457I II87 3372	-I2	5076 7464 4644 6566	-I0	500I 2204 6547 2567	-II	2
3	5803 I774 5074 6246	-I0	0000 0000 0000 0000	0	42I0 0673 4354 76I7	-I0	3
4	520I 573I 0565 5757	- 7	5744 8I2I 240I I240	- 7	576I 4732 5265 57I5	-I0	4
5	4247 I720 0604 66I6	- 6	0000 0000 0000 0000	0	7477 3627 0I77 I3I5	-I0	5
6	6340 7422 4050 6I0I	- 6	45I3 6432 3360 5323	- 6	45I0 0420 0II5 5326	- 7	6
7	4867 7465 600I 247I	- 5	0000 0000 0000 0000	0	5363 403I 5574 57I6	- 7	7
I0	575I 5026 58I5 5242	- 5	6225 5563 2248 6236	- 6	6280 II24 25I6 6425	- 7	I0
II	7508 I2II 7573 7II5	- 5	0000 0000 0000 0000	0	7062 608I 6I67 7757	- 7	II
I2	460I 2855 5445 6750	- 4	77I2 I486 62I0 3455	- 6	77I0 3644 7I5I 3444	- 7	I2
I3	5628 4433 7I00 7I50	- 4	0000 0000 0000 0000	0	4255 3225 3I62 5474	- 6	I3
I4	6727 5275 0034 6462	- 4	456I 7077 4734 5762	- 5	4562 4I72 23I5 2066	- 6	I4
I5	4046 I425 6I00 2230	- 3	0000 0000 0000 0000	0	5062 2747 025I 6676	- 6	I5
I6	4570 204I 4560 I764	- 3	5856 460I 6466 5I53	- 5	5856 063I I636 30I0	- 6	I6
I7	534I 4425 5222 0850	- 3	0000 0000 0000 0000	0	5646 II37 7I37 5532	- 6	I7
20	6I4I 45II II26 6050	- 3	6I30 53II 6I6I 0022	- 5	6I3I 04I5 5207 5655	- 6	20
2I	6767 4507 5400 4653	- 3	0000 0000 0000 0000	0	6405 7453 I402 0404	- 6	2I
22	7642 657I 8525 0I40	- 3	6655 56I0 5022 760I	- 5	6655 3I73 4267 5772	- 6	22
23	426I 2673 7523 4052	- 2	0000 0000 0000 0000	0	7II7 52I7 2I4I 2666	- 6	23
24	4633 I774 6077 4622	- 2	7853 3555 4I25 43I4	- 5	7853 56I2 7423 7072	- 6	24
25	52I6 5337 3545 5857	- 2	0000 0000 0000 0000	0	7600 7244 5665 700I	- 6	25
26	56I3 I645 I306 0070	- 2	4007 7I2I 4205 0534	- 4	4007 6242 0042 II42	- 5	26
27	6220 3746 634I I000	- 2	0000 0000 0000 0000	0	4II3 72I3 3636 2424	- 5	27
30	6636 0346 4023 400I	- 2	42I4 3004 3302 2743	- 4	42I4 3545 7365 5260	- 5	30
3I	7263 3462 6064 7226	- 2	0000 0000 0000 0000	0	43II I472 264I 2463	- 5	3I
32	7720 I473 2652 I344	- 2	4402 2264 I4I0 0247	- 4	4402 I624 I855 I20I	- 5	32
33	4I7I 782I 243I 5II4	- I	0000 0000 0000 0000	0	4467 4300 2I43 4542	- 5	33
34	44I6 7544 2475 I546	- I	4550 6722 4477 0I42	- 4	4550 7272 I072 5774	- 5	34
35	4646 7507 77I4 6476	- I	0000 0000 0000 0000	0	4626 I300 67I3 04I0	- 5	35
36	5I0I 5I4I 6477 4673	- I	4677 3I63 5464 5020	- 4	4677 2675 70I7 3I04	- 5	36
37	5836 6245 3750 3402	- .I	0000 0000 0000 0000	0	4744 3660 6777 0II6	- 5	37
40	5576 0756 3457 2306	- I	5005 2540 5075 47I3	- 4	5005 275I 45I6 I474	- 5	40
4I	6037 2760 I045 4040	- I	0000 0000 0000 0000	0	504I 7300 4766 7205	- 5	4I
42	6302 2I27 5726 270I	- I	5072 I447 II3I 2232	- 4	5072 I307 4722 7545	- 5	42
43	6546 483I 4I32 65I6	- I	0000 0000 0000 0000	0	5II6 0723 5330 7436	- 5	43
44	70I3 743I 7374 752I	- I	5I35 5046 2445 42I5	- 4	5I35 5I36 6032 3473	- 5	44
45	7262 I25I 7I62 I432	- I	0000 0000 0000 0000	0	5I50 57I6 423I 2476	- 5	45
46	7530 7432 607I 3I32	- I	5I57 3280 I5I0 220I	- 4	5I57 3205 4233 I546	- 5	46
47	4000 0000 0000 0000	0	0000 0000 0000 0000	0	5I6I 5075 5373 7200	- 5	47

No.	Nodes	Weights of Gaussian quadratures	Weights of improved quadratures	No.
40	0,5208 9257 3952 289I	0,0407 625I 4640 I929	0,0203 8I22 2549 6278	40
4I	5407 5II6 2085 5522	0000 0000 0000 0000	0203 3033 6945 9389	4I
42	56I0 420I 2668 9337	0404 9I24 6885 2986	0202 4572 8560 645I	42
43	0,58I2 3I3I 0302 743I	0,0000 0000 0000 0000	0,020I 272I 389I 5306	43
44	60I2 8522 6946 0584	0399 505I 662I 7639	0I99 7508 0367 I860	44
45	62II 705I 6784 2755	0000 0000 0000 0000	0I97 9005 4527 6332	45
46	6408 5440 4895 0826	039I 4392 2329 I055	0I95 722I 6238 0672	46
47	6603 0390 5777 7054	0000 0000 0000 0000	0I93 2I29 4268 I47I	47
48	0,6794 8622 0239 7I75	0,0380 7683 I774 2232	0,0I90 3807 72I8 0068	48
49	6983 6974 I8I2 I457	0000 0000 0000 0000	0I87 2384 7087 6I48	49
50	7I69 2358 47I6 I882	0367 5634 6292 37I7	0I83 7860 4463 0093	50
5I	0,735I I644 2478 8755	0000 0000 0000 0000	0I80 0I9I 4535 30I0	5I
52	7529 I735 8963 9656	035I 9I25 3533 4495	0I75-9509 03I2 696I	52
53	0,7702 9724 7723 9682	0,0000 0000 0000 0000	0,0I7I 6004 750I II82	53
54	7872 280I 0523 9035	0333 9I96 8989 5702	0I66 9664 2963 I739	54
55	8036 8079 5357 I952	0000 0000 0000 0000	0I62 04I4 0I04 6360	55
56	8I96 2720 79I4 8408	03I3 7046 6696 0665	0I56 8442 957I 6605	56
57	8350 4I53 099I 7I68	0000 0000 0000 0000	0I5I 402I 0028 6476	57
58	0,8498 9934 0I89 5922	0,029I 40I9 9573 4986	0,0I45 7I08 086I 9445	58
59	864I 7490 3066 63I3	0000 0000 0000 0000	0I39 7575 3703 8706	59
60	8778 4295 I876 9853	0267 I600 9955 I662	0I33 5680 0698 5453	60
6I	8908 8I95 7950 8340	0000 0000 0000 0000	0I27 I800 2486 9775	6I
62	9032 7208 3802 6584	024I I403 0930 3793	0I20 5850 9I95 35I2	62
63	0,9I49 9I2I 7399 6987	0,0000 0000 0000 0000	0,0II3 7599 3895 5656	63
64	9260 I75I 0966 I8II	02I3 5I57 9252 3372	0I06 7390 4070 7855	64
65	9363 3444 78I8 2859	0000 0000 0000 0000	0099 5776 7I30 3I29	65
66	9459 2786 9502 3I6I	0I84 4704 0797 0I24	0092 2595 9237 7702	66
67	9547 7952 I037 47I2	0000 0000 0000 0000	0084 740I 2280 8I76	67
68	0,9628 7066 6024 2922	0,0I54 I975 0272 5875	0,0077 0659 9729 4820	68
69	970I 9074 4907 I34I	0000 0000 0000 0000	0069 3274 7447 5853	69
70	9767 33I6 5466 7648	0I22 8986 9869 II62	006I 4956 9680 7944	70
7I	9824 8328 6875 6282	0000 0000 0000 0000	0053 4726 3867 9452	7I
72	9874 23I6 4295 0768	0090 7828 8854 8066	0045 3I98 7020 2635	72
73	0,99I5 4953 6497 3942	0,0000 0000 0000 0000	0,0037 2264 3254 I033	73
74	9948 6972 7I33 I928	0058 0672 2358 2343	0029 I596 9739 9869	74
75	9973 7238 9622 0696	0000 0000 0000 0000	0020 8292 7980 8920	75
76	9990 2496 5267 8438	0025 0I44 0374 8I97	00I2 2I3I 4635 7948	76
77	9998 38I9 559I 4385	0000 0000 0000 0000	0004 359I 50I9 7892	77
Remainder terms		$G_{75} = 0,I59 \times 10^{-20}$	$K_{II6} = 0,52I \times 10^{-36}$	

No.	Nodes Mantissa	Order	Weights of Gaussian quadratures Mantissa	Order	Weights of improved quadratures Mantissa	Order	No.
50	4123 4162 4743 2322	0	5I57 3230 I5I0 220I	− 4	5I57 3205 4233 I546	− 5	50
5I	4246 7258 0806 7I63	0	0000 0000 0000 0000	0	5I50 57I6 428I 2476	− 5	5I
52	4372 0I63 020I 4I30	0	5I35 5046 2445 42I5	− 4	5I35 5I36 6032 3473	− 5	52
53	45I4 5623 I722 4530	0	0000 0000 0000 0000	0	5II6 0723 5330 7437	− 5	53
54	4636 6724 I024 6437	0	5072 I447 II3I 2233	− 4	5072 I307 4722 7545	− 5	54
55	4760 2407 7355 I760	0	0000 0000 0000 0000	0	504I 7300 4766 7205	− 5	55
56	5I00 74I0 6I50 2634	0	5005 2540 5075 47I3	− 4	5005 275I 45I6 I474	− 5	56
57	5220 4655 20I3 6I77	0	0000 0000 0000 0000	0	4744 3660 6777 0II6	− 5	57
60	5337 I3I7 0540 I442	0	4677 3I63 5464 5020	− 4	4677 2675 70I7 3I04	− 5	60
6I	5454 4I34 003I 4540	0	0000 0000 0000 0000	0	4626 I300 67I3 04I0	− 5	6I
62	5570 4II5 654I 3II5	0	4550 6722 4477 0I42	− 4	4550 7272 I072 5774	− 5	62
63	5703 0227 2563 I332	0	0000 0000 0000 0000	0	4467 4300 2I43 4542	− 5	63
64	60I3 746I I225 3507	0	4402 2264 I4I0 0250	− 4	4402 I624 I355 I20I	− 5	64
65	6I23 I063 2362 6I32	0	0000 0000 0000 0000	0	43II I472 264I 2463	− 5	65
66	6230 3706 2773 I000	0	42I4 3004 3302 2743	− 4	42I4 3545 7365 5260	− 5	66
67	6333 7006 2307 5600	0	0000 0000 0000 0000	0	4II3 72I3 3636 2424	− 5	67
70	6435 I426 55I6 3762	0	4007 7I2I 4205 0534	− 4	4007 6242 0042 II42	− 5	70
7I	6534 25I0 I046 4504	0	0000 0000 0000 0000	0	7600 7244 5665 700I	− 6	7I
72	668I I400 6360 0633	0	7353 3555 4I25 43I4	− 5	7353 56I2 7423 7072	− 6	72
73	6723 522I 0053 0765	0	0000 0000 0000 0000	0	7II7 52I7 2I4I 2666	− 6	73
74	70I3 5I20 6425 2764	0	6655 56I0 5022 760I	− 5	6655 3I73 4267 5772	− 6	74
75	7I0I 0327 0237 73I2	0	0000 0000 0000 0000	0	6405 7453 I402 0404	− 6	75
76	7I63 6326 6665 II73	0	6I30 58II 6I6I 0022	− 5	6I3I 04I5 5207 5655	− 6	76
77	7243 6335 2255 5743	0	0000 0000 0000 0000	0	5646 II37 7I37 5532	− 6	77
I00	7320 7573 632I 760I	0	5356 460I 6466 5I53	− 5	5356 063I I636 30I0	− 6	I00
I0I	7373 I635 2I67 7555	0	0000 0000 0000 0000	0	5062 2747 025I 6676	− 6	I0I
I02	7442 4I24 I376 I455	0	456I 7077 4734 5762	− 5	4562 4I72 23I5 2066	− 6	I02
I03	7506 6I56 2033 748I	0	0000 0000 0000 0000	0	4255 3225 3I62 5474	− 6	I03
I04	7547 726I III5 504I	0	77I2 I436 62I0 3455	− 6	77I0 3644 7I5I 3444	− 7	I04
I05	7605 7I53 5404 I0I5	0	0000 0000 0000 0000	0	7062 603I 6I67 7756	− 7	I05
I06	7640 5457 225I 4453	0	6225 5563 2243 6236	− 6	6230 II24 25I6 6425	− 7	I06
I07	7670 2006 2437 7526	0	0000 0000 0000 0000	0	5363 408I 5574 57I6	− 7	I07
II0	77I4 3703 5537 27I7	0	45I3 6432 3360 5323	− 6	45I0 0420 0II5 5326	− 7	II0
III	7735 3060 577I 73I2	0	0000 0000 0000 0000	0	7477 3627 0I77 I3I5	−I0	III
II2	7752 77I0 2335 05I0	0	5744 3I2I 240I I240	− 7	576I 4732 5265 57I5	−I0	II2
II3	7765 I7I4 0065 6063	0	0000 0000 0000 0000	− 0	42I0 0673 4354 76I7	−I0	II3
II4	7774 003I 5033 3202	0	5076 7464 4644 6566	−I0	500I 2204 6547 2567	−II	II4
II5	7777 2545 2765 5534	0	0000 0000 0000 0000	0	7II0 5606 0I75 40II	−I3	II5

$$n = 39$$

No.	Nodes	Weights of Gaussian quadratures	Weights of improved quadratures	No.
I	0,000I 5374 8232 5I74	0,0000 0000 0000 0000	0,0004 I4I7 0880 2539	I
2	0009 2630 8466 7835	0023 7647 2345 8I76	00II 6020 035I 4667	2
3	0024 9628 8320 I825	0000 0000 0000 0000	00I9 7896 3599 9274	3
4	0048 7423 I572 6570	0055 I739 4469 5823	0027 7082 3376 2286	4
5	0080 2927 I540 9009	0000 0000 0000 0000	0035 3756 5I78 7930	5
6	0,0II9 5064 5333 2645	0,0086 28II 4546 8625	0,0043 0706 8442 III6	6
7	0I66 4582 8II5 0692	0000 0000 0000 0000	0050 8294 5680 225I	7
8	022I I239 3837 6739	0II6 8469 24I6 089I	0058 4697 I063 7026	8
9	0283 3850 090I I3I7	0000 0000 0000 0000	0065 9285 9I59 2929	9
I0	0352 9542 5756 6309	0I46 6747 799I 95I7	0073 3037 4589 8752	I0
II	0,0429 9275 5739 09I4	0,0000 0000 0000 0000	0,0080 6287 3378 7I86	II
I2	05I4 I644 0353 5036	0I75 5755 5749 0657	0087 8I37 5799 96I0	I2
I3	0605 4903 4533 9592	0000 0000 0000 0000	0094 8088 7III 0327	I3
I4	0703 7353 I000 0469	0203 3663 8423 9669	0I0I 662I 8849 84I5	I4
I5	0808 776I 5628 8073	0000 0000 0000 0000	0I08 3962 5990 I556	I5
I6	0,0920 4685 I284 9284	0,0229 87I5 0554 4583	0,0II4 9533 0744 4532	I6
I7	I038 6095 8690 4320	0000 0000 0000 0000	0I2I 2938 2I60 0267	I7
I8	II62 9937 8534 4682	0254 9233 2646 0647	0I27 4465 786I 7900	I8
I9	I293 4464 2452 6823	0000 0000 0000 0000	0I33 4277 6463 5725	I9
20	I429 7778 2052 7327	0278 3634 5I70 458I	0I39 I949 7348 0567	20
2I	0,I57I 7534 5336 6032	0,0000 0000 0000 0000	0,0I44 7I57 5603 6634	2I
22	I7I9 I339 3283 9945	0300 0436 8044 298I	0I50 0099 80I9 2630	22
23	I87I 7029 0802 6705	0000 0000 0000 0000	0I55 0906 I0II 4958	23
24	2029 2327 252I 36I0	03I9 8269 4069 34I2	0I59 9242 5908 0604	24
25	2I9I 4593 3437 4988	0000 0000 0000 0000	0I64 4835 5202 9878	25
26	0,2358 II36 5669 78I3	0,0337 588I 5483 II56	0,0I68 784I I8I2 705I	26
27	2528 9458 0963 2527	0000 0000 0000 0000	0I72 8376 4776 4590	27
28	2703 6974 3845 4320	0353 2I50 2985 3044	0I76 6I68 263I 6670	28
29	2882 0796 2976 05I0	0000 0000 0000 0000	0I80 098I 7679 6776	29
30	3063 799I 80I4 2I93	0366 6087 6707 I343	0I83 2955 7202 8358	30
3I	0,3248 5805 4024 80I2	0,0000 0000 0000 0000	0,0I86 2207 II40 3922	3I
32	3436 I422 0375 9070	0377 6846 866I 4I80	0I88 8507 7273 3929	32
33	3626 I764 3796 735I	0000 0000 0000 0000	0I9I I654 0956 7759	33
34	38I8 3724 3769 082I	0386 3727 6272 34I0	0I93 I782 3I5I 9I53	34
35	40I2 4369 8262 6340	0000 0000 0000 0000	0I94 90I8 585I 6973	35
36	0,4208 0733 000I 08II	0,0392 6I80 6643 6856	0,0I96 3I69 9032 7504	36
37	4404 96I7 I972 4389	0000 0000 0000 0000	0I97 4058 2909 906I	37
38	4602 7809 7695 6223	0396 38II I284 I842	0I98 I827 I080 6255	38
39	480I 2283 66I7 97I6	0000 0000 0000 0000	0I98 66I8 8050 4967	39
40	0,5000 0000 0000 0000	0,0397 638I I069 72I4	0,0I98 8268 6347 494I	40

$$n = 47$$

No.	Nodes Mantissa	Order	Weights of Gaussian quadratures Mantissa	Order	Weights of improved quadratures Mantissa	Order	No.
I	5023 3572 0547 II34	-I4	0000 0000 0000 0000	0	6622 240I 7I74 0700	-I8	I
2	7455 I602 22I0 544I	-I2	4673 7227 0025 0203	-I0	460I 0735 I037 4626	-II	2
3	507I 4304 5445 II62	-I0	0000 0000 0000 0000	0	4033 0577 742I 03I7	-I0	3
4	4773 4004 4624 6774	- 7	55I4 5502 3I62 5226	- 7	558I 3263 62I2 65I2	-I0	4
5	4070 6464 5267 I246	- 6	0000 0000 0000 0000	0	7I75 3I76 600I 4462	-I0	5
6	6074 6243 5575 3I73	- 6	4325 6357 3470 I676	- 6	4322 III7 03I5 6507	- 7	6
7	4205 6324 7274 7526	- 5	0000 0000 0000 0000	0	5I50 7326 5666 0563	- 7	7
I0	5522 24I5 0274 7603	- 5	5767 0446 745I 7I66	- 6	577I 3762 5776 37II	- 7	I0
II	720I 5650 3600 6254	- 5	0000 0000 0000 0000	0	6600 435I I7I7 35I4	- 7	II
I2	44II 0757 5522 5605	- 4	7404 7734 445I I452	- 6	7403 I643 44I4 I300	- 7	I2
I3	540I 4453 7735 2072	- 4	0000 0000 0000 0000	0	4I0I 5044 2744 576I	- 6	I3
I4	645I 50I3 47I3 4I63	- 4	4375 2334 7572 66II	- 5	4375 7702 3577 5542	- 6	I4
I5	7600 II03 5440 5455	- 4	0000 0000 0000 0000	0	4665 267I 0I20 6623	- 6	I5
I6	440I 7777 3423 0646	- 3	5I5I 4405 4742 0343	- 5	5I5I 0066 2757 5243	- 6	I6
I7	5I32 I45I 64I5 2537	- 3	0000 0000 0000 0000	0	543I 4257 6203 582I	- 6	I7
20	57I0 I4I7 I773 7623	- 3	5704 76I4 3670 I0I7	- 5	5705 335I 3I52 50I7	- 6	20
2I	65I3 24I5 7I64 0275	- 3	0000 0000 0000 0000	0	6I53 5I20 73I2 60I3	- 6	2I
22	7342 7I36 2654 705I	- 3	64I5 25I3 7006 7553	- 5	64I4 7370 I334 2050	- 6	22
23	4I07 I354 0II7 6403	- 2	0000 0000 0000 0000	0	665I 565I I062 0343	- 6	23
24	4446 4304 4I76 II50	- 2	7I00 44I4 0II3 32I7	- 5	7I00 7232 36I3 7422	- 6	24
25	50I7 I222 6775 2466	- 2	0000 0000 0000 0000	0	732I 5057 770I 675I	- 6	25
26	5400 5020 4I7I 4270	- 2	7534 5670 337I 4330	- 5	7534 3276 763I 545I	- 6	26
27	5772 462I 456I 740I	- 2	0000 0000 0000 0000	0	774I 4667 3530 0672	- 6	27
30	6374 5436 2304 7I40	- 2	4060 0III 0420 6575	- 4	4060 I2I4 2204 4I06	- 5	30
3I	7006 37I2 5II2 4553	- 2	0000 0000 0000 0000	0	4I53 7263 3526 7462	- 5	3I
32	7427 42I0 6744 7246	- 2	4244 3257 0I63 5024	- 4	4244 2230 2574 7300	- 5	32
33	4027 5546 0I22 2505	- I	0000 0000 0000 0000	0	433I 3256 4356 73I0	- 5	33
34	4246 6747 I364 4675	- I	44I2 6507 6052 6736	- 4	44I2 7473 5227 I556	- 5	34
35	4470 7776 3755 5633	- I	0000 0000 0000 0000	0	4470 4523 I74I 67I5	- 5	35
36	47I5 6724 I000 5542	- I	4542 4667 II77 75I5	- 4	4542 3736 057I 0033	- 5	36
37	5I45 I7I3 3674 0747	- I	0000 0000 0000 0000	0	46I0 6520 2420 2I02	- 5	37
40	5376 7063 7736 4375	- I	4653 I434 0670 4I34	- 4	4653 2340 52I3 0546	- 5	40
4I	5632 4405 044I 7I75	- I	0000 0000 0000 0000	0	47II 5II2 6I6I 523I	- 5	4I
42	6070 0053 65I4 2026	- I	4744 I036 6777 0205	- 4	4744 0I5I 2525 5475	- 5	42
43	6326 7720 3070 3470	- I	0000 0000 0000 0000	0	4772 474I 6646 7233	- 5	43
44	6567 20I6 74I0 5487	- I	50I5 0374 7604 I632	- 4	50I5 I250 I287 0I0I	- 5	44
45	7030 4266 6624 2406	- I	0000 0000 0000 0000	0	5033 3400 600I 0223	- 5	45
46	7272 4622 I025 2750	- I	5045 5622 3644 0242	- 4	5045 4755 2II7 I02I	- 5	46
47	7585 I25I 0452 6203	- I	0000 0000 0000 0000	0	5053 7I52 3762 I542	- 5	47
50	4000 0000 0000 0000	0	5055 7540 6033 66I4	- 4	5056 0403 7474 7705	- 5	50

No.	Nodes	Weights of Gaussian quadratures	Weights of improved quadratures	No.
4I	0,5I98 7716 3882 0284	0,0000 0000 0000 0000	0,0I98 6618 8050 4967	4I
42	5397 2I90 2304 3777	0896 38II I284 I842	0I98 I827 I080 6255	42
43	5595 0382 8027 56II	0000 0000 0000 0000	0I97 4058 2909 906I	43
44	579I 9266 9998 9I89	0392 6I80 6643 6856	0I96 3I69 9032 7504	44
45	0,5987 5630 I737 3660	0,0000 0000 0000 0000	0,0I94 90I8 585I 6973	45
46	6I8I 6275 6230 9I79	0386 3727 6272 34I0	0I93 I782 3I5I 9I53	46
47	6373 8235 6203 2649	0000 0000 0000 0000	0I9I I654 0956 7759	47
48	6563 8577 9624 0930	0377 6846 866I 4I80	0I88 8507 7273 3929	48
49	675I 4I94 5975 I988	0000 0000 0000 0000	0I86 2207 II40 3922	49
50	0,6936 2008 I985 7807	0,0366 6087 6707 I343	0,0I83 2955 7202 8358	50
5I	7II7 9203 7023 9490	0000 0000 0000 0000	0I80 098I 7679 6776	5I
52	7296 3025 6I54 5680	0353 2I50 2985 8044	0I76 6I68 263I 6670	52
53	747I 054I 9036 7473	0000 0000 0000 0000	0I72 8376 4776 4590	53
54	764I 8863 4330 2I87	0337 588I 5483 II56	0I68 784I I8I2 705I	54
55	0,7808 5406 6562 50I2	0,0000 0000 0000 0000	0,0I64 4835 5202 9878	55
56	7970 7672 7478 6390	03I9 8269 4069 34I2	0I59 9242 5908 0604	56
57	8I28 2970 9I97 3295	0000 0000 0000 0000	0I55 0906 I0II 4958	57
58	8280 8660 67I6 0055	0300 0436 8044 298I	0I50 0099 80I9 2630	58
59	8428 2465 4663 3968	0000 0000 0000 0000	0I44 7I57 5603 6634	59
60	0,8570 222I 7947 2673	0,0278 3634 5I70 458I	0,0I39 I949 7348 0567	60
6I	8706 5535 7547 3I77	0000 0000 0000 0000	0I33 4277 6463 5725	6I
62	8837 0062 I465 53I7	0254 9233 2646 0647	0I27 4465 786I 7900	62
63	896I 3904 I309 5680	0000 0000 0000 0000	0I2I 2938 2I60 0267	63
64	9079 53I4 87I5 07I6	0229 87I5 0554 4583	0II4 9533 0744 4532	64
65	0,9I9I 2238 487I I927	0,0000 0000 0000 0000	0,0I08 3962 5990 I556	65
66	9296 2646 8999 953I	0203 3663 8423 9669	0I0I 662I 8849 84I5	66
67	9394 5096 5466 0408	0000 0000 0000 0000	0094 8088 7III 0327	67
68	9485 8355 9646 4964	0I75 5755 5749 0657	0087 8I37 5799 96I0	68
69	9570 0724 4260 9086	0000 0000 0000 0000	0080 6287 3378 7I86	69
70	0,9647 0457 4243 369I	0,0I46 6747 799I 95I7	0,0073 3037 4589 8752	70
7I	97I6 6649 9098 8683	0000 0000 0000 0000	0065 9285 9I59 2929	7I
72	9778 8760 6I62 326I	0II6 8469 24I6 089I	0058 4697 I063 7026	72
73	9833 54I7 I884 9308	0000 0000 0000 0000	0050 8294 5680 225I	73
74	9880 4935 4666 7355	0086 28II 4546 8625	0043 0706 8442 III6	74
75	0,99I9 7072 8459 099I	0,0000 0000 0000 0000	0,0035 3756 5I78 7930	75
76	995I 2576 8427 3430	0055 I739 4469 5823	0027 7082 3376 2286	76
77	9975 037I I679 8I75	0000 0000 0000 0000	00I9 7896 3599 9274	77
78	9990 7369 I583 2I65	0028 7647 2345 8I76	00II 6020 035I 4667	78
79	9998 4625 I767 4826	0000 0000 0000 0000	0004 I4I7 0380 2539	79
Remainder terms		$G_{78} = 0{,}408 \times 10^{-2I}$	$K_{I20} = 0{,}683 \times 10^{-37}$	

No.	Nodes		Weights of Gaussian quadratures		Weights of improved quadratures		No.
	Mantissa	Order	Mantissa	Order	Mantissa	Order	
5I	4I2I 3253 3552 4676	0	0000 0000 0000 0000	0	5053 7I52 3762 I542	− 5	5I
52	4242 5466 7365 24I3	0	5045 5622 3644 0242	− 4	5045 4755 2II7 I02I	− 5	52
53	4363 5644 4465 6574	0	0000 0000 0000 0000	0	5033 3400 600I 0223	− 5	53
54	4504 2770 4I73 5I60	0	50I5 0374 7604 I632	− 4	50I5 I250 I237 0I0I	− 5	54
55	4624 4027 6343 6I43	0	0000 0000 0000 0000	0	4772 474I 6646 7233	− 5	55
56	4743 7752 053I 6764	0	4744 I036 6777 0205	− 4	4744 0I5I 2525 5475	− 5	56
57	5062 5575 3557 030I	0	0000 0000 0000 0000	0	47II 5II2 6I6I 523I	− 5	57
60	5200 4346 0020 560I	0	4653 I434 0670 4I34	− 4	4653 2340 52I3 0546	− 5	60
6I	53I5 3032 204I 74I4	0	0000 0000 0000 0000	0	46I0 6520 2420 2I02	− 5	6I
62	543I 0425 7377 5II6	0	4542 4667 II77 75I5	− 4	4542 3736 057I 0038	− 5	62
63	5543 4000 60II I062	0	0000 0000 0000 0000	0	4470 4523 I74I 67I5	− 5	63
64	5654 44I4 3205 544I	0	44I2 6507 6052 6736	− 4	44I2 7473 5227 I556	− 5	64
65	5764 III4 7726 6585	0	0000 0000 0000 0000	0	433I 3256 4356 73I0	− 5	65
66	6072 0735 6206 6I26	0	4244 3257 0I63 5024	− 4	4244 2230 2574 7300	− 5	66
67	6I76 30I5 2555 2645	0	0000 0000 0000 0000	0	4I53 7263 3526 7462	− 5	67
70	6300 6470 33I6 6I47	0	4060 0III 0420 6575	− 4	4060 I2I4 2204 4I06	− 5	70
7I	640I 2633 4648 4077	0	0000 0000 0000 0000	0	774I 4667 3530 0672	− 6	7I
72	6477 6573 674I 472I	0	7534 5670 337I 4330	− 5	7534 3276 763I 545I	− 6	72
73	6574 I588 2200 5262	0	0000 0000 0000 0000	0	732I 5057 770I 675I	− 6	73
74	6666 27I6 6740 3545	0	7I00 44I4 0II3 32I7	− 5	7I00 7232 36I3 7422	− 6	74
75	6756 I504 7754 0277	0	0000 0000 0000 0000	0	665I 565I I062 0343	− 6	75
76	7043 5064 I5I2 3072	0	64I5 25I3 7006 7553	− 5	64I4 7370 I334 2050	− 6	76
77	7I26 4586 206I 3750	0	0000 0000 0000 0000	0	6I53 5I20 78I2 60I3	− 6	77
I00	7206 7636 0600 40I5	0	5704 76I4 3670 I0I7	− 5	5705 835I 3I52 50I7	− 6	I00
I0I	7264 5632 6I36 2524	0	0000 0000 0000 0000	0	548I 4257 6203 582I	− 6	I0I
I02	7387 6000 0435 47I3	0	5I5I 4405 4742 0343	− 5	5I5I 0066 2757 5243	− 6	I02
I03	7407 7733 6II5 75I5	0	0000 0000 0000 0000	0	4665 267I 0I20 6623	− 6	I03
I04	7455 3I37 2I43 2I70	0	4375 2334 7572 66II	− 5	4375 7702 3577 5542	− 6	I04
I05	75I7 7I55 2002 I274	0	0000 0000 0000 0000	0	4I0I 5044 2744 576I	− 6	I05
I06	7557 334I 0II2 6507	0	7404 7734 445I I452	− 6	7403 I648 44I4 I300	− 7	I06
I07	76I3 7442 5703 7632	0	0000 0000 0000 0000	0	6600 435I I7I7 35I4	− 7	I07
II0	7645 3327 4572 0608	0	5767 0446 745I 7I66	− 6	577I 3762 5776 37II	− 7	II0
III	7673 643I 26I2 0605	0	0000 0000 0000 0000	0	5I50 7326 5666 0563	− 7	III
II2	77I7 03I5 3422 0246	0	4325 6357 3470 I676	− 6	4322 III7 03I5 6507	− 7	II2
II3	7737 07I3 I325 I065	0	0000 0000 0000 0000	0	7I75 3I76 600I 4462	−I0	II3
II4	7754 02I7 7554 6544	0	55I4 5502 3I62 5226	− 7	558I 3263 62I2 65I2	−I0	II4
II5	7765 6I47 I664 6655	0	0000 0000 0000 0000	0	4033 0577 742I 03I7	−I0	II5
II6	7774 I5I3 0766 6735	0	4673 7227 0025 0203	−I0	460I 0735 I037 4626	−II	II6
II7	7777 2754 4205 7230	0	0000 0000 0000 0000	0	6622 240I 7I74 0700	−I3	II7

n = 40

No.	Nodes	Weights of Gaussian quadratures	Weights of improved quadratures	No.
I	0,000I 4622 0370 6500	0,0000 0000 0000 0000	0,0003 9393 I66I 9472	I
2	0008 8II4 5I44 7204	0022 6063 8549 2666	00II 0374 2867 8634	2
3	0023 747I 3276 9636	0000 0000 0000 0000	00I8 826I 4339 67I0	3
4	0046 3688 0650 27I5	0052 49I4 2265 5764	0026 3597 I357 4427	4
5	0076 3858 0067 8750	0000 0000 0000 0000	0033 6590 6742 6037	5
6	0,0II3 7002 5008 II29	0,0082 I052 9I90 9539	0,0040 9878 8I93 3757	6
7	0I58 3848 6572 9250	0000 0000 0000 0000	0048 3770 0742 0086	7
8	02I0 4I59 0393 I042	0III 2292 4597 0835	0055 6566 0832 0I38	8
9	0269 6408 I4I8 7500	0000 0000 0000 0000	0062 77I9 2384 2586	9
I0	0335 9359 5860 66I7	0I39 6850 3490 0II7	0069 8I27 9933 4903	I0
II	0,0409 2522 8463 5506	0,0000 0000 0000 0000	0,0076 8066 3I79 55I2	II
I2	0489 5059 65I5 5629	0I67 3009 764I 2739	0083 6726 6237 50I3	I2
I3	0576 5399 5649 455I	0000 0000 0000 0000	0090 3693 4204 409I	I3
I4	0670 2024 8393 8702	0I93 9I08 3987 2360	0096 9382 2947 I589	I4
I5	0770 3800 7205 3446	0000 0000 0000 0000	0I03 3952 I636 764I	I5
I6	0,0876 9388 4583 3442	0,02I9 3545 4092 8360	0,0I09 6909 3667 9I65	I6
I7	0989 697I 6929 8739	0000 0000 0000 0000	0II5 7946 5506 6885	I7
I8	II08 47I7 4286 7403	0243 4790 38I7 536I	0I2I 7284 509I I367	I8
I9	I233 I009 8280 5289	0000 0000 0000 0000	0I27 50I0 880I 565I	I9
20	I363 4087 2405 0364	0266 I392 349I 9684	0I33 0786 8749 5I23	20
2I	0,I499 I85I I256 3350	0,0000 0000 0000 0000	0,0I38 438I 3055 5305	2I
22	I640 2I65 7692 9I02	0287 I988 4549 6958	0I43 59I9 3420 546I	22
23	I786 3023 7847 2I00	0000 0000 0000 0000	0I48 5446 3638 8883	23
24	I937 2305 5I66 0099	0306 58I2 I246 4645	0I53 27I8 0445 7058	24
25	2092 7646 7085 4850	0000 0000 0000 0000	0I57 756I I809 5577	25
26	0,2252 6643 7452 4359	0,0324 0200 6728 3005	0,0I62 0049 I253 8030	26
27	24I6 6969 6306 808I	0000 0000 0000 0000	0I66 0202 2I70 6288	27
28	2584 6209 9I56 9I06	0339 5602 2907 6I70	0I69 7843 I4I7 I049	28
29	2756 I774 3I80 9I8I	0000 0000 0000 0000	0I73 2846 792I 7488	29
30	293I I039 78I4 I975	0353 0582 3695 6434	0I76 5257 2354 3I09	30
3I	0,3I09 I428 2263 2055	0,0000 0000 0000 0000	0,0I79 5080 I39I 8I4I	3I
32	3290 0295 4587 I208	0364 4329 II97 9020	0I82 2I9I 3265 I705	32
33	3473 4877 9I32 3766	0000 0000 0000 0000	0I84 6508 4767 0243	33
34	3659 2390 7496 3732	0373 6I58 4528 984I	0I86 8059 00I2 7346	34
35	3847 0073 9059 6402	0000 0000 0000 0000	0I88 6840 063I 5468	35
36	0,4036 5I20 9649 3I44	0,0380 55I8 0950 3I3I	0,0I90 2773 I889 4262	36
37	4227 4656 03I0 3028	0000 0000 0000 0000	0I9I 58I6 2002 5874	37
38	44I9 5796 4662 3724	0385 I990 9082 I240	0I92 5987 0874 9754	38
39	46I2 5670 5834 3586	0000 0000 0000 0000	0I93 3277 77I9 5705	39
40	4806 I379 I246 9746	0387 5297 3989 2I24	0I93 765I 4689 3762	40
4I	0,5000 0000 0000 0000	0,0000 0000 0000 0000	0,0I93 9I05 2382 I4I4	4I

$$n = 50$$

No.	Nodes Mantissa	Order	Weights of Gaussian quadratures Mantissa	Order	Weights of improved quadratures Mantissa	Order	No.
I	4625 I273 2677 25I0	-I4	0000 0000 0000 0000	0	6350 4234 6556 0545	-I3	I
2	7I57 6246 2430 I607	-I2	4502 3470 506I 7006	-I0	44I2 5567 025I 23I4	-II	2
3	4672 0423 67I2 3722	-I0	0000 0000 0000 0000	0	7554 I0I6 0242 6020	-II	3
4	4577 037I 2774 524I	- 7	5300 0377 I346 6405	- 7	53I4 0I02 6I37 7047	-I0	4
5	7644 64I5 6242 47I3	- 7	0000 0000 0000 0000	0	67II 32I3 3030 0236	-I0	5
6	5644 4527 3I02 06II	- 6	4I50 2564 4556 4573	- 6	4I44 7440 4763 2054	- 7	6
7	4033 7633 7063 345I	- 5	0000 0000 0000 0000	0	4750 2623 I7I4 2377	- 7	7
I0	5305 755I 6460 I705	- 5	5543 6355 2605 3635	- 6	5546 0045 50I0 3475	- 7	I0
II	67I6 I706 7670 3334	- 5	0000 0000 0000 0000	0	6333 0350 0002 I70I	- 7	II
I2	423I 4560 I623 4657	- 4	7II5 6046 3247 4560	- 6	7II4 I470 3775 0340	- 7	I2
I3	5I72 0466 2725 0402	- 4	0000 0000 0000 0000	0	7672 7022 4527 45II	- 7	I3
I4	62I0 0I53 5447 3354	- 4	4220 66I6 6I64 6422	- 5	422I 3333 5543 2442	- 6	I4
I5	7302 3230 47I4 2456	- 4	0000 0000 0000 0000	0	4500 7646 I564 6I62	- 6	I5
I6	4224 075I 3602 5770	- 3	4755 50I4 7436 2264	- 5	4755 I327 I040 006I	- 6	I6
I7	4734 3032 2300 02I2	- 3	0000 0000 0000 0000	0	5226 3430 6445 0340	- 6	I7
20	547I 433I 7454 0736	- 3	5473 0773 I747 I733	- 5	5473 3666 5II0 II64	- 6	20
2I	6253 0242 430I 0304	- 3	0000 0000 0000 0000	0	5733 37I4 3240 I077	- 6	2I
22	7060 I727 7065 I422	- 3	6I67 250I 2476 55I0	- 5	6I67 0234 7I03 22I0	- 6	22
23	77I0 500I I660 6444	- 3	0000 0000 0000 0000	0	64I6 2725 05I2 7267	- 6	23
24	427I 636I 0405 4730	- 2	6640 256I 30I0 2440	- 5	6640 4477 2074 0I35	- 6	24
25	4630 2074 7672 2324	- 2	0000 0000 0000 0000	0	7055 045I 05I5 20I0	- 6	25
26	5I77 25I3 0750 07I4	- 2	7264 2766 4705 48I3	- 5	7264 I322 42I5 4000	- 6	26
27	5556 5330 2550 263I	- 2	0000 0000 0000 0000	0	7466 0042 7035 2530	- 6	27
30	6I45 7526 I232 6507	- 2	766I 6I0I I2I4 I660	- 5	766I 7332 5404 2303	- 6	30
3I	6544 622I 7575 0504	- 2	0000 0000 0000 0000	0	4023 5733 0627 6I63	- 5	3I
32	7I52 6076 5557 5203	- 2	4II3 3767 347I 0256	- 4	4II3 3344 4072 3654	- 5	32
33	7567 4I02 6204 4670	- 2	0000 0000 0000 0000	0	4200 0366 62I2 3I6I	- 5	33
34	4I05 2444 7344 2325	- I	426I 2570 3743 7500	- 4	426I 3I3I 758I 0042	- 5	34
35	432I 6704 6445 063I	- I	0000 0000 0000 0000	0	4337 2I56 3007 0062	- 5	35
36	454I I220 7227 2604	- I	44II 6326 704I 4434	- 4	44II 6040 4747 3432	- 5	36
37	4763 0050 II05 60I4	- I	0000 0000 0000 0000	0	4460 66I7 0724 3II7	- 5	37
40	5207 I423 2I07 5I46	- I	4524 26I7 3732 4I22	- 4	4524 3037 I426 6056	- 5	40
4I	5435 3662 7725 566I	- I	0000 0000 0000 0000	0	4564 2026 7I42 33I6	- 5	4I
42	5665 5I40 6760 5374	- I	4620 4I65 7655 6020	- 4	4620 40II 3I3I 337I	- 5	42
43	6II7 3576 6240 206I	- I	0000 0000 0000 0000	0	465I 0747 3507 6576	- 5	43
44	6352 5537 04I5 72II	- I	4675 7677 6637 7722	- 4	4676 00I3 6I35 2033	- 5	44
45	6607 I074 5540 7370	- I	0000 0000 0000 0000	0	47I7 0623 720I 2450	- 5	45
46	7044 4I20 36I5 6374	- I	4734 34I5 2700 5267	- 4	4734 3340 3420 7775	- 5	46
47	7302 4726 6I23 6420	- I	0000 0000 0000 0000	0	4745 7706 0027 60I4	- 5	47
50	754I I402 6I55 I434	- I	4753 5560 I750 0243	- 4	4753 5577 07I5 3673	- 5	50
5I	4000 0000 0000 0000	0	0000 0000 0000 0000	0	4755 4773 764I 75I3	- 5	5I

No.	Nodes	Weights of Gaussian quadratures	Weights of improved quadratures	No.
42	0,5193 8620 8753 0254	0,0387 5297 3989 2I24	0,0I93 765I 4689 3762	42
43	5387 4329 4I65 64I4	0000 0000 0000 0000	0I93 3277 77I9 5705	43
44	5580 4203 5337 6276	0385 I990 9082 I240	0I92 5987 0874 9754	44
45	5772 5343 9689 6972	0000 0000 0000 0000	0I9I 58I6 2002 5874	45
46	5963 4879 0350 6855	0380 55I8 0950 3I3I	0I90 2773 I889 4262	46
47	0,6I52 9926 0940 3597	0,0000 0000 0000 0000	0,0I88 6840 063I 5468	47
48	6340 7609 2503 6268	0373 6I58 4528 984I	0I86 8059 00I2 7346	48
49	6526 5I22 0867 6234	0000 0000 0000 0000	0I84 6508 4767 0248	49
50	6709 9704 54I2 8792	0364 4329 II97 9020	0I82 2I9I 3265 I705	50
5I	6890 857I 7736 7945	0000 0000 0000 0000	0I79 5080 I39I 8I4I	5I
52	0,7068 8960 2I85 8025	0,0353 0582 3695 6484	0,0I76 5257 2354 3I09	52
53	7243 8225 68I9 08I9	0000 0000 0000 0000	0I73 2846 792I 7488	53
54	74I5 3790 0843 0894	0339 5602 2907 6I70	0I69 7843 I4I7 I049	54
55	7583 3030 3693 I9I9	0000 0000 0000 0000	0I66 0202 2I70 6288	55
56	7747 3356 2547 564I	0324 0200 6728 3005	0I62 0049 I253 8030	56
57	0,7907 2353 29I4 5650	0,0000 0000 0000 0000	0,0I57 756I I809 5577	57
58	8062 7694 4833 990I	0306 58I2 I246 4645	0I53 27I8 0445 7058	58
59	82I3 6976 2I52 7900	0000 0000 0000 0000	0I48 5446 3688 8883	59
60	8359 7834 2307 0898	0287 I988 4549 6958	0I43 59I9 3420 546I	60
6I	8500 8I48 8743 6650	0000 0000 0000 0000	0I38 438I 3055 5305	6I
62	0,8636 59I2 7594 9636	0,0266 I392 349I 9684	0,0I33 0786 8749 5I23	62
63	8766 8990 I7I9 47II	0000 0000 0000 0000	0I27 50I0 880I 565I	63
64	889I 5282 57I3 2597	0243 4790 38I7 536I	0I2I 7284 509I I367	64
65	90I0 3028 3070 I26I	0000 0000 0000 0000	0II5 7946 5506 6885	65
66	9I23 06II 54I6 6558	02I9 3545 4092 8366	0I09 6909 3667 9I65	66
67	0,9229 6I99 2793 6554	0,0000 0000 0000 0000	0,0I03 3952 I636 764I	67
68	9329 7975 I606 I298	0I93 9I08 3987 2360	0096 9382 2947 I589	68
69	9423 4600 4350 5449	0000 0000 0000 0000	0090 3693 4204 409I	69
70	95I0 4940 3484 437I	0I67 3009 764I 2739	0083 6726 6287 50I3	70
7I	9590 7477 I536 4494	0000 0000 0000 0000	0076 8066 3I79 55I2	7I
72	0,9664 0640 4I39 3383	0,0I39 6850 3490 0II7	0,0069 8I27 9933 4903	72
73	9730 359I 858I 2500	0000 0000 0000 0000	0062 77I9 2384 2586	73
74	9789 5840 9606 8958	0III 2292 4597 0835	0055 6566 0832 0I38	74
75	984I 6I56 3427 0750	0000 0000 0000 0000	0048 3770 0742 0086	75
76	9886 2997 499I 887I	0082 I052 9I90 9539	0040 9878 8I93 3757	76
77	0,9923 6I4I 9932 I250	0,0000 0000 0000 0000	0,0033 6590 6742 6037	77
78	9953 63II 9349 7285	0052 49I4 2265 5764	0026 3597 I357 4427	78
79	9976 2528 6723 0364	0000 0000 0000 0000	00I8 826I 4339 67I0	79
80	999I I885 4855 2796	0022 6063 8549 2666	00II 0374 2867 8634	80
8I	9998 5377 9629 3500	0000 0000 0000 0000	0003 9393 I66I 9472	8I
Remainder terms		$G_{80} = 0,104 \times 10^{-2I}$	$K_{122} = 0,659 \times 10^{-38}$	

No.	Nodes		Weights of Gaussian quadratures		Weights of improved quadratures		No.
	Mantissa	Order	Mantissa	Order	Mantissa	Order	
52	4II7 3I76 47II 3I6I	0	4753 5560 I750 0243	- 4	4753 5577 07I5 3673	- 5	52
53	4236 5424 4726 0570	0	0000 0000 0000 0000	0	4745 7706 0027 60I4	- 5	53
54	4355 5727 607I 0602	0	4734 34I5 2700 5267	- 4	4734 3340 3420 7775	- 5	54
55	4474 334I 5II7 4203	0	0000 0000 0000 0000	0	47I7 0623 720I 2450	- 5	55
56	46I2 5I20 357I 0273	0	4675 7677 6637 7722	- 4	4676 00I3 6I35 2033	- 5	56
57	4730 2I00 4657 6747	0	0000 0000 0000 0000	0	465I 0747 3507 6576	- 5	57
60	5045 I3I7 4407 5202	0	4620 4I65 7655 6020	- 4	4620 40II 3I3I 337I	- 5	60
6I	5I6I 2046 4025 I047	0	0000 0000 0000 0000	0	4564 2026 7I42 33I6	- 5	6I
62	5274 3I66 2734 I3I5	0	4524 26I7 3732 4I22	- 4	4524 3037 I426 6056	- 5	62
63	5406 3753 7335 0772	0	0000 0000·0000 0000	0	4460 66I7 0724 3II7	- 5	63
64	55I7 3267 4264 2475	0	44II 6326 704I 4434	- 4	44II 6040 4747 3453	- 5	64
65	5627 0435 4555 3463	0	0000 0000 0000 0000	0	4337 2I56 3007 0062	- 5	65
66	5735 2555 42I5 6625	0	426I 2570 3748 7500	- 4	426I 3I3I 753I 0042	- 5	66
67	6042 0757 2336 6622	0	0000 0000 0000 0000	0	4200 0366 62I2 3I60	- 5	67
70	6I45 2360 2444 0537	0	4II3 3767 347I 0256	- 4	4II3 3844 4072 3654	- 5	70
7I	6246 6333 4040 5657	0	0000 0000 0000 0000	0	4023 5733 0627 6I63	- 5	7I
72	6346 4052 353I 2256	0	766I 6I0I I2I4 I660	- 5	766I 7332 5404 2303	- 6	72
73	6444 25II 7245 723I	0	0000 0000 0000 0000	0	7466 0042 7035 2530	- 6	73
74	6540 I255 I605 76I5	0	7264 2766 4705 48I3	- 5	7264 I322 42I5 4000	- 6	74
75	663I 7360 602I 33I3	0	0000 0000 0000 0000	0	7055 045I 05I5 20I0	- 6	75
76	672I 4303 5676 46I2	0	6640 256I 30I0 2440	- 5	6640 4477 2074 0I35	- 6	76
77	7006 7277 66II 7I33	0	0000 0000 0000 0000	0	64I6 2725 05I2 7267	- 6	77
I00	707I 7605 007I 2636	0	6I67 250I 2476 55I0	- 5	6I67 0234 7I03 22I0	- 6	I00
I0I	7I52 4753 5347 6747	0	0000 0000 0000 0000	0	5733 37I4 3240 I077	- 6	I0I
I02	7230 6344 6032 3704	0	5473 0773 I747 I733	- 5	5473 3666 5II0 II64	- 6	I02
I03	7304 3474 5547 7757	0	0000 0000 0000 0000	0	5226 3430 6445 0340	- 6	I03
I04	7355 3702 64I7 520I	0	4755 50I4 7436 2264	- 5	4755 I327 I040 006I	- 6	I04
I05	7423 6626 3543 I655	0	0000 0000 0000 0000	0	4500 7646 I564 6I62	- 6	I05
I06	7467 377I 2II5 422I	0	4220 66I6 6I64 6422	- 5	422I 3333 5543 2442	- 6	I06
I07	7530 2754 4642 5360	0	0000 0000 0000 0000	0	7672 7022 4527 45II	- 7	I07
II0	7566 3I50 7706 6I45	0	7II5 6046 3247 4560	- 6	7II4 I470 3775 0340	- 7	II0
III	762I 434I 6202 I7II	0	0000 0000 0000 0000	0	6333 0350 0002 I700	- 7	III
II2	765I 6404 5426 374I	0	5543 6355 2605 3635	- 6	5546 0045 50I0 3475	- 7	II2
II3	7677 I003 I0I6 3I07	0	0000 0000 0000 0000	0	4750 2623 I7I4 2377	- 7	II3
II4	772I 3332 5046 7572	0	4I50 2564 4556 4573	- 6	4I44 7440 4763 2054	- 7	II4
II5	7740 5545 7I06 5654	0	0000 0000 0000 0000	0	67II 32I3 3080 0236	-I0	II5
II6	7755 0036 0324 0I53	0	5300 0377 I346 6405	- 7	53I4 0I02 6I37 7046	-I0	II6
II7	7766 2I36 7302 I530	0	0000 0000 0000 0000	0	7554 I0I6 0242 6020	-II	II7
I20	7774 3I00 6546 5637	0	4502 3470 506I 7006	-I0	44I2 5567 025I 23I4	-II	I20
I2I	7777 3I52 6504 5I0I	0	0000 0000 0000 0000	0	6350 4234 6556 0545	-I3	I2I

TABLES OF
RELATIVE ERRORS IN
QUADRATURE FORMULAS

Relative Error in Gaussian Quadratures

k \ n	1	2	3	4	5
2	$1{,}000$				
4		$0{,}444.10^0$			
6		$0{,}741.10^0$	$0{,}160.10^0$		
8		$0{,}889.10^0$	$0{,}352.10^0$	$0{,}522.10^{-1}$	
10		$0{,}955.10^0$	$0{,}525.10^0$	$0{,}142.10^0$	$0{,}161.10^{-1}$
12			$0{,}663.10^0$	$0{,}248.10^0$	$0{,}520.10^{-1}$
14			$0{,}767.10^0$	$0{,}357.10^0$	$0{,}104.10^0$
16			$0{,}841.10^0$	$0{,}459.10^0$	$0{,}167.10^0$
18			$0{,}894.10^0$	$0{,}552.10^0$	$0{,}236.10^0$
20			$0{,}929.10^0$	$0{,}633.10^0$	$0{,}306.10^0$
22				$0{,}702.10^0$	$0{,}376.10^0$
24				$0{,}760.10^0$	$0{,}443.10^0$
26				$0{,}807.10^0$	$0{,}506.10^0$
28				$0{,}847.10^0$	$0{,}564.10^0$
30				$0{,}878.10^0$	$0{,}618.10^0$
32				$0{,}904.10^0$	$0{,}666.10^0$
34					$0{,}709.10^0$
36					$0{,}747.10^0$
38					$0{,}781.10^0$
40					$0{,}811.10^0$
42					$0{,}837.10^0$
44					$0{,}860.10^0$
46					$0{,}880.10^0$
48					$0{,}897.10^0$
50					$0{,}912.10^0$

Relative Error in Gaussian Quadratures

k \ n	6	7	8	9	10
12	$0,480.10^{-2}$				
14	$0,179.10^{-1}$	$0,139.10^{-2}$			
16	$0,403.10^{-1}$	$0,588.10^{-2}$	$0,396.10^{-3}$		
18	$0,718.10^{-1}$	$0,147.10^{-1}$	$0,187.10^{-2}$	$0,111.10^{-3}$	
20	$0,109.10^{0}$	$0,285.10^{-1}$	$0,516.10^{-2}$	$0,581.10^{-3}$	$0,307.10^{-4}$
22	$0,153.10^{0}$	$0,472.10^{-1}$	$0,108.10^{-1}$	$0,174.10^{-2}$	$0,176.10^{-3}$
24	$0,200.10^{0}$	$0,705.10^{-1}$	$0,193.10^{-1}$	$0,395.10^{-2}$	$0,573.10^{-3}$
26	$0,249.10^{0}$	$0,978.10^{-1}$	$0,306.10^{-1}$	$0,752.10^{-2}$	$0,139.10^{-2}$
28	$0,298.10^{0}$	$0,128.10^{0}$	$0,449.10^{-1}$	$0,127.10^{-1}$	$0,282.10^{-2}$
30	$0,348.10^{0}$	$0,161.10^{0}$	$0,619.10^{-1}$	$0,196.10^{-1}$	$0,504.10^{-2}$
32	$0,397.10^{0}$	$0,196.10^{0}$	$0,814.10^{-1}$	$0,283.10^{-1}$	$0,819.10^{-2}$
34	$0,443.10^{0}$	$0,232.10^{0}$	$0,103.10^{0}$	$0,389.10^{-1}$	$0,124.10^{-1}$
36	$0,488.10^{0}$	$0,269.10^{0}$	$0,127.10^{0}$	$0,513.10^{-1}$	$0,178.10^{-1}$
38	$0,531.10^{0}$	$0,306.10^{0}$	$0,152.10^{0}$	$0,653.10^{-1}$	$0,243.10^{-1}$
40	$0,571.10^{0}$	$0,343.10^{0}$	$0,178.10^{0}$	$0,809.10^{-1}$	$0,321.10^{-1}$
42	$0,609.10^{0}$	$0,379.10^{0}$	$0,206.10^{0}$	$0,979.10^{-1}$	$0,411.10^{-1}$
44	$0,644.10^{0}$	$0,415.10^{0}$	$0,234.10^{0}$	$0,116.10^{0}$	$0,512.10^{-1}$
46	$0,677.10^{0}$	$0,449.10^{0}$	$0,262.10^{0}$	$0,136.10^{0}$	$0,625.10^{-1}$
48	$0,707.10^{0}$	$0,483.10^{0}$	$0,291.10^{0}$	$0,156.10^{0}$	$0,748.10^{-1}$
50	$0,735.10^{0}$	$0,515.10^{0}$	$0,319.10^{0}$	$0,177.10^{0}$	$0,881.10^{-1}$
52	$0,761.10^{0}$	$0,546.10^{0}$	$0,348.10^{0}$	$0,198.10^{0}$	$0,102.10^{0}$
54	$0,784.10^{0}$	$0,576.10^{0}$	$0,376.10^{0}$	$0,220.10^{0}$	$0,117.10^{0}$
56	$0,805.10^{0}$	$0,604.10^{0}$	$0,403.10^{0}$	$0,243.10^{0}$	$0,133.10^{0}$
58	$0,825.10^{0}$	$0,631.10^{0}$	$0,430.10^{0}$	$0,266.10^{0}$	$0,149.10^{0}$
60	$0,843.10^{0}$	$0,656.10^{0}$	$0,457.10^{0}$	$0,288.10^{0}$	$0,166.10^{0}$
62	$0,859.10^{0}$	$0,680.10^{0}$	$0,483.10^{0}$	$0,311.10^{0}$	$0,183.10^{0}$
64	$0,873.10^{0}$	$0,703.10^{0}$	$0,508.10^{0}$	$0,334.10^{0}$	$0,201.10^{0}$

Relative Error in Gaussian Quadratures

k \ n	I1	I2	I3	I4	I5
22	$0,843.10^{-5}$				
24	$0,526.10^{-4}$	$0,229.10^{-5}$			
26	$0,184.10^{-3}$	$0,155.10^{-4}$	$0,620.10^{-6}$		
28	$0,478.10^{-3}$	$0,580.10^{-4}$	$0,449.10^{-5}$	$0,167.10^{-6}$	
30	$0,103.10^{-2}$	$0,160.10^{-3}$	$0,180.10^{-4}$	$0,129.10^{-5}$	$0,446.10^{-7}$
32	$0,193.10^{-2}$	$0,364.10^{-3}$	$0,527.10^{-4}$	$0,549.10^{-5}$	$0,368.10^{-6}$
34	$0,330.10^{-2}$	$0,721.10^{-3}$	$0,126.10^{-3}$	$0,170.10^{-4}$	$0,166.10^{-5}$
36	$0,523.10^{-2}$	$0,129.10^{-2}$	$0,262.10^{-3}$	$0,428.10^{-4}$	$0,540.10^{-5}$
38	$0,779.10^{-2}$	$0,213.10^{-2}$	$0,490.10^{-3}$	$0,932.10^{-4}$	$0,143.10^{-4}$
40	$0,111.10^{-1}$	$0,330.10^{-2}$	$0,843.10^{-3}$	$0,182.10^{-3}$	$0,325.10^{-4}$
42	$0,151.10^{-1}$	$0,486.10^{-2}$	$0,136.10^{-2}$	$0,325.10^{-3}$	$0,660.10^{-4}$
44	$0,200.10^{-1}$	$0,687.10^{-2}$	$0,207.10^{-2}$	$0,543.10^{-3}$	$0,123.10^{-3}$
46	$0,256.10^{-1}$	$0,936.10^{-2}$	$0,302.10^{-2}$	$0,858.10^{-3}$	$0,212.10^{-3}$
48	$0,322.10^{-1}$	$0,124.10^{-1}$	$0,424.10^{-2}$	$0,129.10^{-2}$	$0.346.10^{-3}$
50	$0,395.10^{-1}$	$0,159.10^{-1}$	$0,577.10^{-2}$	$0,187.10^{-2}$	$0,538.10^{-3}$
52	$0,477.10^{-1}$	$0,201.10^{-1}$	$0,764.10^{-2}$	$0,261.10^{-2}$	$0,802.10^{-3}$
54	$0,566.10^{-1}$	$0,248.10^{-1}$	$0,987.10^{-2}$	$0,355.10^{-2}$	$0,115.10^{-2}$
56	$0,663.10^{-1}$	$0,301.10^{-1}$	$0,125.10^{-1}$	$0,470.10^{-2}$	$0,161.10^{-2}$
58	$0,767.10^{-1}$	$0,360.10^{-1}$	$0,155.10^{-1}$	$0.609.10^{-2}$	$0,218.10^{-2}$
60	$0,877.10^{-1}$	$0,425.10^{-1}$	$0,189.10^{-1}$	$0,773.10^{-2}$	$0,289.10^{-2}$
62	$0,994.10^{-1}$	$0,496.10^{-1}$	$0,228.10^{-1}$	$0,963.10^{-2}$	$0,374.10^{-2}$
64	$0,112.10^{0}$	$0,572.10^{-1}$	$0,271.10^{-1}$	$0,118.10^{-1}$	$0,477.10^{-2}$

Relative Error in Gaussian Quadratures

k \ n	I6	I7	I8	I9	20
32	$0,119.10^{-7}$				
34	$0,104.10^{-6}$	$0,815.10^{-8}$			
36	$0,494.10^{-6}$	$0,292.10^{-7}$	$0,834.10^{-9}$		
38	$0,169.10^{-5}$	$0,146.10^{-6}$	$0,812.10^{-8}$	$0,220.10^{-9}$	
40	$0,469.10^{-5}$	$0,524.10^{-6}$	$0,427.10^{-7}$	$0,225.10^{-8}$	$0,579.10^{-10}$
42	$0,111.10^{-4}$	$0,152.10^{-5}$	$0,161.10^{-6}$	$0,124.10^{-7}$	$0,622.10^{-9}$
44	$0,235.10^{-4}$	$0,375.10^{-5}$	$0,484.10^{-6}$	$0,487.10^{-7}$	$0,358.10^{-8}$
46	$0,453.10^{-4}$	$0,823.10^{-5}$	$0,125.10^{-5}$	$0,158.10^{-6}$	$0,146.10^{-7}$
48	$0,812.10^{-4}$	$0,164.10^{-4}$	$0,284.10^{-5}$	$0,408.10^{-6}$	$0,478.10^{-7}$
50	$0,136.10^{-3}$	$0,305.10^{-4}$	$0,587.10^{-5}$	$0,964.10^{-6}$	$0,132.10^{-6}$
52	$0,219.10^{-3}$	$0,529.10^{-4}$	$0,112.10^{-4}$	$0,206.10^{-5}$	$0,323.10^{-6}$
54	$0,336.10^{-3}$	$0,873.10^{-4}$	$0,201.10^{-4}$	$0,407.10^{-5}$	$0,714.10^{-6}$
56	$0,495.10^{-3}$	$0,138.10^{-3}$	$0,341.10^{-4}$	$0,751.10^{-5}$	$0,145.10^{-5}$
58	$0,708.10^{-3}$	$0,208.10^{-3}$	$0,552.10^{-4}$	$0,131.10^{-4}$	$0,276.10^{-5}$
60	$0,984.10^{-3}$	$0,305.10^{-3}$	$0,859.10^{-4}$	$0,218.10^{-4}$	$0,495.10^{-5}$
62	$0,133.10^{-2}$	$0,434.10^{-3}$	$0,129.10^{-3}$	$0,347.10^{-4}$	$0,844.10^{-5}$
64	$0,177.10^{-2}$	$0,602.10^{-3}$	$0,188.10^{-3}$	$0,534.10^{-4}$	$0,138.10^{-4}$
66	$0,230.10^{-2}$	$0,815.10^{-3}$	$0,266.10^{-3}$	$0,795.10^{-4}$	$0,217.10^{-4}$
68	$0,293.10^{-2}$	$0,108.10^{-2}$	$0,367.10^{-3}$	$0,115.10^{-3}$	$0,330.10^{-4}$
70	$0,368.10^{-2}$	$0,141.10^{-2}$	$0,497.10^{-3}$	$0,162.10^{-3}$	$0,489.10^{-4}$
72	$0,456.10^{-2}$	$0,180.10^{-2}$	$0,659.10^{-3}$	$0,224.10^{-3}$	$0,703.10^{-4}$
74	$0,558.10^{-2}$	$0,227.10^{-2}$	$0,859.10^{-3}$	$0,303.10^{-3}$	$0,989.10^{-4}$
76	$0,674.10^{-2}$	$0,282.10^{-2}$	$0,110.10^{-2}$	$0,401.10^{-3}$	$0,136.10^{-3}$
78	$0,805.10^{-2}$	$0,346.10^{-2}$	$0,139.10^{-2}$	$0,524.10^{-3}$	$0,184.10^{-3}$
80	$0,952.10^{-2}$	$0,420.10^{-2}$	$0,174.10^{-2}$	$0,673.10^{-3}$	$0,244.10^{-3}$
82	$0,112.10^{-1}$	$0,504.10^{-2}$	$0,214.10^{-2}$	$0,852.10^{-3}$	$0,319.10^{-3}$
84	$0,130.10^{-1}$	$0,599.10^{-2}$	$0,261.10^{-2}$	$0,107.10^{-2}$	$0,410.10^{-3}$
86	$0,150.10^{-1}$	$0,706.10^{-2}$	$0,314.10^{-2}$	$0,132.10^{-2}$	$0,521.10^{-3}$
88	$0,171.10^{-1}$	$0,825.10^{-2}$	$0,375.10^{-2}$	$0,161.10^{-2}$	$0,658.10^{-3}$
90	$0,195.10^{-1}$	$0,957.10^{-2}$	$0,444.10^{-2}$	$0,195.10^{-2}$	$0,811.10^{-3}$
92	$0,221.10^{-1}$	$0,110.10^{-1}$	$0,522.10^{-2}$	$0,234.10^{-2}$	$0,995.10^{-3}$
94	$0,248.10^{-1}$	$0,126.10^{-1}$	$0,608.10^{-2}$	$0,278.10^{-2}$	$0,121.10^{-2}$
96	$0,277.10^{-1}$	$0,143.10^{-1}$	$0,704.10^{-2}$	$0,328.10^{-2}$	$0,146.10^{-2}$

Relative Error in Gaussian Quadratures

k \ n	21	22	23	24	25
42	$0{,}152.10^{-10}$				
44	$0{,}171.10^{-9}$	$0{,}397.10^{-11}$			
46	$0{,}102.10^{-8}$	$0{,}467.10^{-10}$	$0{,}104.10^{-11}$		
48	$0{,}436.10^{-8}$	$0{,}292.10^{-9}$	$0{,}127.10^{-10}$	$0{,}271.10^{-12}$	
50	$0{,}148.10^{-7}$	$0{,}129.10^{-8}$	$0{,}826.10^{-10}$	$0{,}345.10^{-11}$	$0{,}705.10^{-13}$
52	$0{,}424.10^{-7}$	$0{,}454.10^{-8}$	$0{,}379.10^{-9}$	$0{,}233.10^{-10}$	$0{,}933.10^{-12}$
54	$0{,}107.10^{-6}$	$0{,}185.10^{-7}$	$0{,}138.10^{-8}$	$0{,}111.10^{-9}$	$0{,}653.10^{-11}$
56	$0{,}244.10^{-6}$	$0{,}351.10^{-7}$	$0{,}423.10^{-8}$	$0{,}417.10^{-9}$	$0{,}322.10^{-10}$
58	$0{,}511.10^{-6}$	$0{,}825.10^{-7}$	$0{,}114.10^{-7}$	$0{,}132.10^{-8}$	$0{,}125.10^{-9}$
60	$0{,}999.10^{-6}$	$0{,}178.10^{-6}$	$0{,}276.10^{-7}$	$0{,}366.10^{-8}$	$0{,}409.10^{-9}$
62	$0{,}184.10^{-5}$	$0{,}857.10^{-6}$	$0{,}611.10^{-7}$	$0{,}911.10^{-8}$	$0{,}117.10^{-8}$
64	$0{,}322.10^{-5}$	$0{,}675.10^{-6}$	$0{,}126.10^{-6}$	$0{,}208.10^{-7}$	$0{,}299.10^{-8}$
66	$0{,}539.10^{-5}$	$0{,}121.10^{-5}$	$0{,}244.10^{-6}$	$0{,}440.10^{-7}$	$0{,}699.10^{-8}$
68	$0{,}868.10^{-5}$	$0{,}208.10^{-5}$	$0{,}449.10^{-6}$	$0{,}874.10^{-7}$	$0{,}152.10^{-7}$
70	$0{,}135.10^{-4}$	$0{,}342.10^{-5}$	$0{,}788.10^{-6}$	$0{,}164.10^{-6}$	$0{,}309.10^{-7}$
72	$0{,}208.10^{-4}$	$0{,}543.10^{-5}$	$0{,}133.10^{-5}$	$0{,}295.10^{-6}$	$0{,}595.10^{-7}$
74	$0{,}300.10^{-4}$	$0{,}838.10^{-5}$	$0{,}215.10^{-5}$	$0{,}508.10^{-6}$	$0{,}109.10^{-6}$
76	$0\ 429\ 10^{-4}$	$0\ 125\ 10^{-4}$	$0\ 339\ 10^{-5}$	$0\ 842\ 10^{-6}$	$0\ 192\ 10^{-6}$
78	$0{,}602.10^{-4}$	$0{,}183.10^{-4}$	$0{,}517.10^{-5}$	$0{,}135.10^{-5}$	$0{,}325.10^{-6}$
80	$0{,}828.10^{-4}$	$0{,}262.10^{-4}$	$0{,}770.10^{-5}$	$0{,}210.10^{-5}$	$0{,}531.10^{-6}$
82	$0{,}112.10^{-3}$	$0{,}366.10^{-4}$	$0{,}112.10^{-4}$	$0{,}319.10^{-5}$	$0{,}843.10^{-6}$
84	$0{,}148.10^{-3}$	$0{,}502.10^{-4}$	$0{,}159.10^{-4}$	$0{,}472.10^{-5}$	$0{,}130.10^{-5}$
86	$0{,}194.10^{-3}$	$0{,}678.10^{-4}$	$0{,}222.10^{-4}$	$0{,}682.10^{-5}$	$0{,}196.10^{-5}$
88	$0{,}250.10^{-3}$	$0{,}900.10^{-4}$	$0{,}305.10^{-4}$	$0{,}968.10^{-5}$	$0{,}288.10^{-5}$
90	$0{,}318.10^{-3}$	$0{,}118.10^{-3}$	$0{,}411.10^{-4}$	$0{,}135.10^{-4}$	$0{,}415.10^{-5}$
92	$0{,}400.10^{-3}$	$0{,}152.10^{-3}$	$0{,}545.10^{-4}$	$0{,}184.10^{-4}$	$0{,}588.10^{-5}$
94	$0{,}497.10^{-3}$	$0{,}194.10^{-3}$	$0{,}714.10^{-4}$	$0{,}249.10^{-4}$	$0{,}816.10^{-5}$
96	$0{,}612.10^{-3}$	$0{,}244.10^{-3}$	$0{,}923.10^{-4}$	$0{,}330.10^{-4}$	$0{,}112.10^{-4}$

Relative Error in Gaussian Quadratures

k \ n	26	27	28	29	30
52	$0,183.10^{-13}$				
54	$0,252.10^{-12}$	$0,475.10^{-14}$			
56	$0,182.10^{-11}$	$0,677.10^{-13}$	$0,123.10^{-14}$		
58	$0,929.10^{-11}$	$0,508.10^{-12}$	$0,182.10^{-13}$	$0\ 319\ 10^{-15}$	
60	$0,373.10^{-10}$	$0,267.10^{-11}$	$0,141.10^{-12}$	$0,486.10^{-14}$	$0,824.10^{-16}$
62	$0,125.10^{-9}$	$0,110.10^{-10}$	$0,764.10^{-12}$	$0,389.10^{-13}$	$0,130.10^{-14}$
64	$0,368.10^{-9}$	$0,383.10^{-10}$	$0,325.10^{-11}$	$0,217.10^{-12}$	$0,107.10^{-13}$
66	$0,970.10^{-9}$	$0,116.10^{-9}$	$0,116.10^{-10}$	$0,953.10^{-12}$	$0,616.10^{-13}$
68	$0,233.10^{-8}$	$0,312.10^{-9}$	$0,359.10^{-10}$	$0,349.10^{-11}$	$0,278.10^{-12}$
70	$0,519.18^{-8}$	$0,769.10^{-9}$	$0,998.10^{-10}$	$0,111.10^{-10}$	$0,105.10^{-11}$
72	$0,108.10^{-7}$	$0,175.10^{-8}$	$0,252.10^{-9}$	$0,316.10^{-10}$	$0,342.10^{-11}$
74	$0,213.10^{-7}$	$0,374.10^{-8}$	$0,588.10^{-9}$	$0,819.10^{-10}$	$0,997.10^{-11}$
76	$0,399.10^{-7}$	$0,754.10^{-8}$	$0,128.10^{-8}$	$0,196.10^{-9}$	$0,264.10^{-10}$
78	$0,717.10^{-7}$	$0,144.10^{-7}$	$0,264.10^{-8}$	$0,437.10^{-9}$	$0,646.10^{-10}$
80	$0,124.10^{-6}$	$0,264.10^{-7}$	$0,517.10^{-8}$	$0,919.10^{-9}$	$0,147.10^{-9}$
82	$0,206.10^{-6}$	$0,466.10^{-7}$	$0,967.10^{-8}$	$0,183.10^{-8}$	$0,316.10^{-9}$
84	$0,334.10^{-6}$	$0,792.10^{-7}$	$0,174.10^{-7}$	$0,350.10^{-8}$	$0,644.10^{-9}$
86	$0,524.10^{-6}$	$0,130.10^{-6}$	$0,301.10^{-7}$	$0,640.10^{-8}$	$0,125.10^{-8}$
88	$0,803.10^{-6}$	$0,208.10^{-6}$	$0,504.10^{-7}$	$0,113.10^{-7}$	$0,233.10^{-8}$
90	$0,120.10^{-5}$	$0,325.10^{-6}$	$0,820.10^{-7}$	$0,192.10^{-7}$	$0,419.10^{-8}$
92	$0,176.10^{-5}$	$0,494.10^{-6}$	$0,130.10^{-6}$	$0,318.10^{-7}$	$0,728.10^{-8}$
94	$0,253.10^{-5}$	$0,735.10^{-6}$	$0,210.10^{-6}$	$0,513.10^{-7}$	$0,122.10^{-7}$
96	$0,356.10^{-5}$	$0,107.10^{-5}$	$0,303.10^{-6}$	$0,806.10^{-7}$	$0,200.10^{-7}$

Relative Error in Gaussian Quadratures

k \ n	31	32	33	34	35
62	$0{,}213.10^{-16}$				
64	$0{,}345.10^{-15}$	$0{,}549.10^{-17}$			
66	$0{,}294.10^{-14}$	$0{,}920.10^{-16}$	$0{,}142.10^{-17}$		
68	$0{,}174.10^{-13}$	$0{,}805.10^{-15}$	$0{,}244.10^{-16}$	$0{,}365.10^{-18}$	
70	$0{,}807.10^{-13}$	$0{,}490.10^{-14}$	$0{,}220.10^{-15}$	$0{,}647.10^{-17}$	$0{,}938.10^{-19}$
72	$0{,}312.10^{-12}$	$0{,}233.10^{-13}$	$0{,}137.10^{-14}$	$0{,}599.10^{-16}$	$0{,}171.10^{-17}$
74	$0{,}104.10^{-11}$	$0{,}924.10^{-13}$	$0{,}672.10^{-14}$	$0{,}384.10^{-15}$	$0{,}163.10^{-16}$
76	$0{,}312.10^{-11}$	$0{,}317.10^{-12}$	$0{,}273.10^{-13}$	$0{,}193.10^{-14}$	$0{,}107.10^{-15}$
78	$0{,}846.10^{-11}$	$0{,}970.10^{-12}$	$0{,}959.10^{-13}$	$0{,}802.10^{-14}$	$0{,}550.10^{-15}$
80	$0{,}211.10^{-10}$	$0.269.10^{-11}$	$0{,}300{,}10^{-12}$	$0{,}288.10^{-13}$	$0{,}234.10^{-14}$
82	$0{,}493.10^{-10}$	$0{,}687.10^{-11}$	$0{,}850.10^{-12}$	$0{,}923.10^{-13}$	$0{,}863.10^{-14}$
84	$0{,}108.10^{-9}$	$0.164.10^{-10}$	$0{,}222.10^{-11}$	$0{,}267.10^{-12}$	$0{,}282.10^{-13}$
86	$0{,}224.10^{-9}$	$0{,}366.10^{-10}$	$0{,}539.10^{-11}$	$0{,}712.10^{-12}$	$0{,}835.10^{-13}$
88	$0{,}445.10^{-9}$	$0.775.10^{-10}$	$0{,}123.10^{-10}$	$0.176.10^{-11}$	$0{,}227.10^{-12}$
90	$0{,}844.10^{-9}$	$0{,}156.10^{-9}$	$0{,}266.10^{-10}$	$0{,}410.10^{-11}$	$0{,}574.10^{-12}$
92	$0{,}154.10^{-8}$	$0{,}802.10^{-9}$	$0{,}546.10^{-10}$	$0{,}903.10^{-11}$	$0{,}136.10^{-11}$
94	$0{,}272.10^{-8}$	$0{,}563.10^{-9}$	$0{,}107.10^{-9}$	$0{,}189{,}10^{-10}$	$0{,}304.10^{-11}$
96	$0{,}466.10^{-8}$	$0{,}101.10^{-8}$	$0{,}203.10^{-9}$	$0{,}379.10^{-10}$	$0{,}650.10^{-11}$
98	$0{,}775.10^{-8}$	$0{,}176.10^{-8}$	$0{,}371.10^{-9}$	$0{,}728.10^{-10}$	$0{,}132.10^{-10}$
100	$0{,}126.10^{-7}$	$0{,}297.10^{-8}$	$0{,}656.10^{-9}$	$0{,}135.10^{-9}$	$0{,}259.10^{-10}$
102	$0{,}199.10^{-7}$	$0{,}489.10^{-8}$	$0{,}113.10^{-8}$	$0{,}243.10^{-9}$	$0{,}489.10^{-10}$
104	$0{,}308.10^{-7}$	$0{,}784.10^{-8}$	$0{,}188.10^{-8}$	$0{,}423.10^{-9}$	$0{,}892.10^{-10}$
106	$0{,}466.10^{-7}$	$0{,}128.10^{-7}$	$0{,}307.10^{-8}$	$0{,}718.10^{-9}$	$0{,}158.10^{-9}$
108	$0{,}693.10^{-7}$	$0{,}189.10^{-7}$	$0{,}488.10^{-8}$	$0{,}119.10^{-8}$	$0{,}272.10^{-9}$
110	$0{,}101.10^{-6}$	$0{,}285.10^{-7}$	$0{,}761.10^{-8}$	$0{,}192.10^{-8}$	$0{,}455.10^{-9}$
112	$0{,}146.10^{-6}$	$0{,}423.10^{-7}$	$0{,}116.10^{-7}$	$0{,}308.10^{-8}$	$0{,}746.10^{-9}$
114	$0{,}206.10^{-6}$	$0{,}616.10^{-7}$	$0{,}174.10^{-7}$	$0{,}469.10^{-8}$	$0{,}119.10^{-8}$
116	$0{,}288.10^{-6}$	$0{,}883.10^{-7}$	$0{,}257.10^{-7}$	$0{,}713.10^{-8}$	$0{,}187.10^{-8}$
118	$0{,}397.10^{-6}$	$0{,}125.10^{-6}$	$0{,}373.10^{-7}$	$0{,}106.10^{-7}$	$0{,}288.10^{-8}$
120	$0{,}540.10^{-6}$	$0{,}174.10^{-6}$	$0{,}534.10^{-7}$	$0{,}156.10^{-7}$	$0{,}436.10^{-8}$
122	$0{,}725.10^{-6}$	$0{,}239.10^{-6}$	$0{,}753.10^{-7}$	$0{,}226.10^{-7}$	$0{,}649.10^{-8}$
124	$0{,}964.10^{-6}$	$0{,}325.10^{-6}$	$0{,}105.10^{-6}$	$0{,}323.10^{-7}$	$0{,}950.10^{-8}$
126	$0{,}127.10^{-5}$	$0{,}487.10^{-6}$	$0{,}144.10^{-6}$	$0{,}454.10^{-7}$	$0{,}137.10^{-7}$
128	$0{,}165.10^{-5}$	$0{,}581.10^{-6}$	$0{,}196.10^{-6}$	$0{,}631.10^{-7}$	$0{,}195.10^{-7}$

n \ k	36	37	38	39	40
72	$0,24I.I0^{-I9}$				
74	$0,452.I0^{-I8}$	$0,620.I0^{-20}$			
76	$0,44I.I0^{-I7}$	$0,II9.I0^{-I8}$	$0,I59.I0^{-20}$		
78	$0,298.I0^{-I6}$	$0,II9.I0^{-I7}$	$0,8I4.I0^{-I9}$	$0,408.I0^{-2I}$	
80	$0,I57.I0^{-I5}$	$0,825.I0^{-I7}$	$0,322.I0^{-I8}$	$0,826.I0^{-20}$	$0,I05.I0^{-2I}$
82	$0,683.I0^{-I5}$	$0,444.I0^{-I6}$	$0,228.I0^{-I7}$	$0,867.I0^{-I9}$	$0,2I7.I0^{-20}$
84	$0,257.I0^{-I4}$	$0,I98.I0^{-I5}$	$0,I26.I0^{-I6}$	$0,629.I0^{-I8}$	$0,233.I0^{-I9}$
86	$0,858.I0^{-I4}$	$0,763.I0^{-I5}$	$0,573.I0^{-I6}$	$0,354.I0^{-I7}$	$0,I73.I0^{-I8}$
88	$0,259.I0^{-I3}$	$0,260.I0^{-I4}$	$0,225.I0^{-I5}$	$0,I65.I0^{-I6}$	$0,997.I0^{-I8}$
90	$0,720.I0^{-I3}$	$0,802.I0^{-I4}$	$0,785.I0^{-I5}$	$0,663.I0^{-I6}$	$0,475.I0^{-I7}$
92	$0,I85.I0^{-I2}$	$0,227.I0^{-I3}$	$0,247.I0^{-I4}$	$0,236.I0^{-I5}$	$0,I94.I0^{-I6}$
94	$0,448.I0^{-I2}$	$0,596.I0^{-I3}$	$0,7I2.I0^{-I4}$	$0,756.I0^{-I5}$	$0,705.I0^{-I6}$
96	$0,I02.I0^{-II}$	$0,I46.I0^{-I2}$	$0,I90.I0^{-I3}$	$0,222.I0^{-I4}$	$0,280.I0^{-I5}$
98	$0,222.I0^{-II}$	$0,34I.I0^{-I2}$	$0,477.I0^{-I3}$	$0,605.I0^{-I4}$	$0,690.I0^{-I5}$
100	$0,460.I0^{-II}$	$0,752.I0^{-I2}$	$0,II3.I0^{-I2}$	$0,I54.I0^{-I3}$	$0,I9I.I0^{-I4}$
102	$0,9I4.I0^{-II}$	$0,I58.I0^{-II}$	$0,253.I0^{-I2}$	$0,37I.I0^{-I3}$	$0,497.I0^{-I4}$
104	$0,I75.I0^{-I0}$	$0,320.I0^{-II}$	$0,542.I0^{-I2}$	$0,847.I0^{-I3}$	$0,I2I.I0^{-I3}$
106	$0,325.I0^{-I0}$	$0,624.I0^{-II}$	$0,III.I0^{-II}$	$0,I84.I0^{-I2}$	$0,282.I0^{-I3}$
108	$0,583.I0^{-I0}$	$0,II7.I0^{-I0}$	$0,220.I0^{-II}$	$0,385.I0^{-I2}$	$0,623.I0^{-I3}$
110	$0,I02.I0^{-9}$	$0,2I4.I0^{-I0}$	$0,42I.I0^{-II}$	$0,773.I0^{-I2}$	$0,I32.I0^{-I2}$
112	$0,I73.I0^{-9}$	$0,379.I0^{-I0}$	$0,778.I0^{-II}$	$0,I50.I0^{-II}$	$0,269.I0^{-I2}$
114	$0,288.I0^{-9}$	$0,653.I0^{-I0}$	$0,I40.I0^{-I0}$	$0,28I.I0^{-II}$	$0,530.I0^{-I2}$
116	$0,467.I0^{-9}$	$0,II0.I0^{-9}$	$0,245.I0^{-I0}$	$0,5I2.I0^{-II}$	$0,I0I.I0^{-II}$
118	$0,742.I0^{-9}$	$0,I8I.I0^{-9}$	$0,4I7.I0^{-I0}$	$0,908.I0^{-II}$	$0,I86.I0^{-II}$
120	$0,II6.I0^{-8}$	$0,29I.I0^{-9}$	$0,695.I0^{-I0}$	$0,I57.I0^{-I0}$	$0,335.I0^{-II}$
122	$0,I77.I0^{-8}$	$0,460.I0^{-9}$	$0,II3.I0^{-9}$	$0,265.I0^{-I0}$	$0\ 586\ I0^{-II}$
124	$0,267.I0^{-8}$	$0,7I2.I0^{-9}$	$0,I8I.I0^{-9}$	$0,438.I0^{-I0}$	$0,I00.I0^{-I0}$
126	$0,394.I0^{-8}$	$0,I09.I0^{-8}$	$0,284.I0^{-9}$	$0,708.I0^{-I0}$	$0,I68.I0^{-I0}$
128	$0,576.I0^{-8}$	$0,I63.I0^{-8}$	$0,438.I0^{-9}$	$0,II2.I0^{-9}$	$0,275.I0^{-I0}$

Relative Error in Improved Quadratures

n / k	I	2	3	4	5
6	$0,160.10^{0}$				
8	$0,352.10^{0}$	$0,163.10^{-I}$			
I0	$0,525.10^{0}$	$0,298.10^{-I}$			
I2	$0,767.10^{0}$	$0,294.10^{-I}$	$0,182.10^{-2}$		
I4	$0,84I.10^{0}$	$0,128.10^{-I}$	$0,56I.10^{-2}$	$0,755.10^{-4}$	
I6	$0,894.10^{0}$	$0,18I.10^{-I}$	$0,102.10^{-I}$	$0,49I.10^{-3}$	
I8	$0,929.10^{0}$	$0,60I.10^{-I}$	$0,139.10^{-I}$	$0,152.10^{-2}$	$0,840.10^{-5}$
20		$0,110.10^{0}$	$0,155.10^{-I}$	$0,330.10^{-2}$	$0,58I.10^{-4}$
22		$0,164.10^{0}$	$0,140.10^{-I}$	$0,577.10^{-2}$	$0,207.10^{-3}$
24		$0,222.10^{0}$	$0,898.10^{-2}$	$0,876.10^{-2}$	$0,52I.10^{-3}$
26		$0,279.10^{0}$	$0,204.10^{-3}$	$0,I20.10^{-I}$	$0,I06.10^{-2}$
28		$0,337.10^{0}$	$0,I22.10^{-I}$	$0,I5I.10^{-I}$	$0,I86.10^{-2}$
30		$0,392.10^{0}$	$0,280.10^{-I}$	$0,I78.10^{-I}$	$0,294.10^{-2}$
32		$0,445.10^{0}$	$0,467.10^{-I}$	$0,I98.10^{-I}$	$0,426.10^{-2}$
34		$0,496.10^{0}$	$0,68I.10^{-I}$	$0,2I0.10^{-I}$	$0,580.10^{-2}$
36		$0,543.10^{0}$	$0,9I7.10^{-I}$	$0,2I0.10^{-I}$	$0,747.10^{-2}$
38		$0,587.10^{0}$	$0,II7.10^{0}$	$0,199.10^{-I}$	$0,92I.10^{-2}$
40		$0,628.10^{0}$	$0,I43.10^{0}$	$0,I75.10^{-I}$	$0,I09.10^{-I}$
42		$0,666.10^{0}$	$0,I72.10^{0}$	$0,I37.10^{-I}$	$0,I25.10^{-I}$
44		$0,700.10^{0}$	$0,20I.10^{0}$	$0,870.10^{-2}$	$0,I40.10^{-I}$
46		$0,732.10^{0}$	$0,230.10^{0}$	$0,236.10^{-2}$	$0,I5I.10^{-I}$
48		$0,760.10^{0}$	$0,259.10^{0}$	$0,522.10^{-2}$	$0,I59.10^{-I}$
50		$0,786.10^{0}$	$0,289.10^{0}$	$0,I40.10^{-I}$	$0,I64.10^{-I}$
52		$0,809.10^{0}$	$0,3I8.10^{0}$	$0,239.10^{-I}$	$0,I64.10^{-I}$
54		$0,830.10^{0}$	$0,347.10^{0}$	$0,348.10^{-I}$	$0,I58.10^{-I}$
56		$0,849.10^{0}$	$0,376.10^{0}$	$0,467.10^{-I}$	$0,I48.10^{-I}$
58		$0,866.10^{0}$	$0,404.10^{0}$	$0,595.10^{-I}$	$0,I32.10^{-I}$
60		$0,882.10^{0}$	$0,432.10^{0}$	$0,730.10^{-I}$	$0,III.10^{-I}$
62		$0,895.10^{0}$	$0,458.10^{0}$	$0,873.10^{-I}$	$0,835.10^{-2}$
64		$0,907.10^{0}$	$0,484.10^{0}$	$0,I02.10^{0}$	$0,506.10^{-2}$
66			$0,5I0.10^{0}$	$0,II8.10^{0}$	$0,II9.10^{-2}$
68			$0,534.10^{0}$	$0,I33.10^{0}$	$0,323.10^{-2}$
70			$0,558.10^{0}$	$0,I50.10^{0}$	$0,82I.10^{-2}$
72			$0,58I.10^{0}$	$0,I66.10^{0}$	$0,I37.10^{-I}$
74			$0,602.10^{0}$	$0,I83.10^{0}$	$0,I98.10^{-I}$

Relative Error in Improved Quadratures

k \ n	6	7	8	9	10
20	$0{,}685 . 10^{-6}$				
22	$0{,}445 . 10^{-5}$				
24	$0{,}180 . 10^{-4}$	$0{,}717 . 10^{-7}$			
26	$0{,}585 . 10^{-4}$	$0{,}550 . 10^{-6}$	$0{,}646 . 10^{-8}$		
28	$0{,}129 . 10^{-3}$	$0{,}238 . 10^{-5}$	$0{,}538 . 10^{-7}$		
30	$0{,}266 . 10^{-3}$	$0{,}756 . 10^{-5}$	$0{,}247 . 10^{-6}$	$0{,}789 . 10^{-9}$	
32	$0{,}489 . 10^{-3}$	$0{,}195 . 10^{-4}$	$0{,}881 . 10^{-6}$	$0{,}668 . 10^{-8}$	$0{,}726 . 10^{-10}$
34	$0{,}820 . 10^{-3}$	$0{,}434 . 10^{-4}$	$0{,}228 . 10^{-5}$	$0{,}330 . 10^{-7}$	$0{,}708 . 10^{-9}$
36	$0{,}128 . 10^{-2}$	$0{,}858 . 10^{-4}$	$0{,}542 . 10^{-5}$	$0{,}118 . 10^{-6}$	$0{,}374 . 10^{-8}$
38	$0{,}187 . 10^{-2}$	$0{,}155 . 10^{-3}$	$0{,}115 . 10^{-4}$	$0{,}343 . 10^{-6}$	$0{,}142 . 10^{-7}$
40	$0{,}261 . 10^{-2}$	$0{,}259 . 10^{-3}$	$0{,}223 . 10^{-4}$	$0{,}856 . 10^{-6}$	$0{,}436 . 10^{-7}$
42	$0{,}349 . 10^{-2}$	$0{,}407 . 10^{-3}$	$0{,}403 . 10^{-4}$	$0{,}191 . 10^{-5}$	$0{,}114 . 10^{-6}$
44	$0{,}451 . 10^{-2}$	$0{,}609 . 10^{-3}$	$0{,}683 . 10^{-4}$	$0{,}388 . 10^{-5}$	$0{,}267 . 10^{-6}$
46	$0{,}564 . 10^{-2}$	$0{,}871 . 10^{-3}$	$0{,}110 . 10^{-3}$	$0{,}733 . 10^{-5}$	$0{,}567 . 10^{-6}$
48	$0{,}687 . 10^{-2}$	$0{,}120 . 10^{-2}$	$0{,}169 . 10^{-3}$	$0{,}130 . 10^{-4}$	$0{,}112 . 10^{-5}$
50	$0{,}818 . 10^{-2}$	$0{,}160 . 10^{-2}$	$0{,}249 . 10^{-3}$	$0{,}218 . 10^{-4}$	$0{,}207 . 10^{-5}$
52	$0{,}953 . 10^{-2}$	$0{,}207 . 10^{-2}$	$0{,}355 . 10^{-3}$	$0{,}350 . 10^{-4}$	$0{,}362 . 10^{-5}$
54	$0{,}109 . 10^{-1}$	$0{,}262 . 10^{-2}$	$0{,}490 . 10^{-3}$	$0{,}540 . 10^{-4}$	$0{,}606 . 10^{-5}$
56	$0{,}122 . 10^{-1}$	$0{,}324 . 10^{-2}$	$0{,}659 . 10^{-3}$	$0{,}803 . 10^{-4}$	$0{,}975 . 10^{-5}$
58	$0{,}136 . 10^{-1}$	$0{,}393 . 10^{-2}$	$0{,}867 . 10^{-3}$	$0{,}116 . 10^{-3}$	$0{,}151 . 10^{-4}$
60	$0{,}148 . 10^{-1}$	$0{,}468 . 10^{-2}$	$0{,}111 . 10^{-2}$	$0{,}162 . 10^{-3}$	$0{,}227 . 10^{-4}$
62	$0{,}159 . 10^{-1}$	$0{,}548 . 10^{-2}$	$0{,}140 . 10^{-2}$	$0{,}222 . 10^{-3}$	$0{,}332 . 10^{-4}$
64	$0{,}169 . 10^{-1}$	$0{,}633 . 10^{-2}$	$0{,}174 . 10^{-2}$	$0{,}296 . 10^{-3}$	$0{,}473 . 10^{-4}$
66	$0{,}177 . 10^{-1}$	$0{,}722 . 10^{-2}$	$0{,}212 . 10^{-2}$	$0{,}388 . 10^{-3}$	$0{,}658 . 10^{-4}$
68	$0{,}184 . 10^{-1}$	$0{,}813 . 10^{-2}$	$0{,}255 . 10^{-2}$	$0{,}499 . 10^{-3}$	$0{,}897 . 10^{-4}$
70	$0{,}188 . 10^{-1}$	$0{,}905 . 10^{-2}$	$0{,}302 . 10^{-2}$	$0{,}681 . 10^{-3}$	$0{,}120 . 10^{-3}$
72	$0{,}190 . 10^{-1}$	$0{,}998 . 10^{-2}$	$0{,}354 . 10^{-2}$	$0{,}785 . 10^{-3}$	$0{,}158 . 10^{-3}$
74	$0{,}190 . 10^{-1}$	$0{,}109 . 10^{-1}$	$0{,}410 . 10^{-2}$	$0{,}964 . 10^{-3}$	$0{,}204 . 10^{-3}$
76	$0{,}187 . 10^{-1}$	$0{,}118 . 10^{-1}$	$0{,}470 . 10^{-2}$	$0{,}117 . 10^{-2}$	$0{,}259 . 10^{-3}$
78	$0{,}182 . 10^{-1}$	$0{,}126 . 10^{-1}$	$0{,}533 . 10^{-2}$	$0{,}140 . 10^{-2}$	$0{,}326 . 10^{-3}$
80	$0{,}174 . 10^{-1}$	$0{,}134 . 10^{-1}$	$0{,}600 . 10^{-2}$	$0{,}166 . 10^{-2}$	$0{,}404 . 10^{-3}$
82	$0{,}163 . 10^{-1}$	$0{,}142 . 10^{-1}$	$0{,}670 . 10^{-2}$	$0{,}194 . 10^{-2}$	$0{,}495 . 10^{-3}$
84	$0{,}149 . 10^{-1}$	$0{,}148 . 10^{-1}$	$0{,}742 . 10^{-2}$	$0{,}226 . 10^{-2}$	$0{,}600 . 10^{-3}$
86	$0{,}132 . 10^{-1}$	$0{,}154 . 10^{-1}$	$0{,}816 . 10^{-2}$	$0{,}260 . 10^{-2}$	$0{,}720 . 10^{-3}$
88	$0{,}113 . 10^{-1}$	$0{,}159 . 10^{-1}$	$0{,}891 . 10^{-2}$	$0{,}297 . 10^{-2}$	$0{,}856 . 10^{-3}$
90	$0{,}902 . 10^{-2}$	$0{,}163 . 10^{-1}$	$0{,}967 . 10^{-2}$	$0{,}337 . 10^{-2}$	$0{,}101 . 10^{-2}$
92	$0{,}648 . 10^{-2}$	$0{,}166 . 10^{-1}$	$0{,}104 . 10^{-1}$	$0{,}379 . 10^{-2}$	$0{,}118 . 10^{-2}$
94	$0{,}364 . 10^{-2}$	$0{,}167 . 10^{-1}$	$0{,}112 . 10^{-1}$	$0{,}424 . 10^{-2}$	$0{,}137 . 10^{-2}$
96	$0{,}523 . 10^{-3}$	$0{,}168 . 10^{-1}$	$0{,}119 . 10^{-1}$	$0{,}470 . 10^{-2}$	$0{,}157 . 10^{-2}$

Relative Error in Improved Quadratures

k \ n	I1	I2	I3	I4	I5
36	$0,840.10^{-11}$				
38	$0,880.10^{-10}$	$0,867.10^{-12}$			
40	$0,496.10^{-9}$	$0,972.10^{-11}$			
42	$0,200.10^{-8}$	$0,584.10^{-10}$	$0,101.10^{-12}$		
44	$0,648.10^{-8}$	$0.250.10^{-9}$	$0,121.10^{-11}$	$0,108.10^{-13}$	
46	$0,179.10^{-7}$	$0,853.10^{-9}$	$0,770.10^{-11}$	$0,137.10^{-12}$	
48	$0,437.10^{-7}$	$0,247.10^{-8}$	$0,348.10^{-10}$	$0,924.10^{-12}$	$0,127.10^{-14}$
50	$0,968.10^{-7}$	$0,632.10^{-8}$	$0,125.10^{-9}$	$0,440.10^{-11}$	$0,170.10^{-13}$
52	$0,199.10^{-6}$	$0,146.10^{-7}$	$0,379.10^{-9}$	$0,166.10^{-10}$	$0,121.10^{-12}$
54	$0,381.10^{-6}$	$0,311.10^{-7}$	$0,101.10^{-8}$	$0,529.10^{-10}$	$0,606.10^{-12}$
56	$0,692.10^{-6}$	$0,619.10^{-7}$	$0,244.10^{-8}$	$0,148.10^{-9}$	$0,240.10^{-11}$
58	$0,120.10^{-5}$	$0,116.10^{-6}$	$0,541.10^{-8}$	$0,371.10^{-9}$	$0,797.10^{-11}$
60	$0,199.10^{-5}$	$0,208.10^{-6}$	$0,112.10^{-7}$	$0,853.10^{-9}$	$0,282.10^{-10}$
62	$0,819.10^{-5}$	$0,357.10^{-6}$	$0,217.10^{-7}$	$0,183.10^{-8}$	$0,606.10^{-10}$
64	$0,494.10^{-5}$	$0,589.10^{-6}$	$0,402.10^{-7}$	$0,368.10^{-8}$	$0,145.10^{-9}$
66	$0,744.10^{-5}$	$0,941.10^{-6}$	$0,710.10^{-7}$	$0,702.10^{-8}$	$0,321.10^{-9}$
68.	$0,109.10^{-4}$	$0,146.10^{-5}$	$0,121.10^{-6}$	$0,128.10^{-7}$	$0,669.10^{-9}$
70	$0,156.10^{-4}$	$0,220.10^{-5}$	$0,198.10^{-6}$	$0,224.10^{-7}$	$0,132.10^{-8}$
72	$0,219.10^{-4}$	$0,324.10^{-5}$	$0,315.10^{-6}$	$0,377.10^{-7}$	$0,248.10^{-8}$
74	$0,801.10^{-4}$	$0,467.10^{-5}$	$0,488.10^{-6}$	$0,616.10^{-7}$	$0,446.10^{-8}$
76	$0,407.10^{-4}$	$0,660.10^{-5}$	$0,737.10^{-6}$	$0,978.10^{-7}$	$0,773.10^{-8}$
78	$0,540.10^{-4}$	$0,916.10^{-5}$	$0,109.10^{-5}$	$0,151.10^{-6}$	$0,180.10^{-7}$
80	$0,706.10^{-4}$	$0,125.10^{-4}$	$0,157.10^{-5}$	$0,228.10^{-6}$	$0,211.10^{-7}$
82	$0,910.10^{-4}$	$0,168.10^{-4}$	$0,223.10^{-5}$	$0,338.10^{-6}$	$0,334.10^{-7}$
84	$0,116.10^{-3}$	$0,222.10^{-4}$	$0,311.10^{-5}$	$0,490.10^{-6}$	$0,516.10^{-7}$
86	$0,145.10^{-3}$	$0,290.10^{-4}$	$0,426.10^{-5}$	$0,697.10^{-6}$	$0,780.10^{-7}$
88	$0,181.10^{-3}$	$0,373.10^{-4}$	$0,576.10^{-5}$	$0,976.10^{-6}$	$0,115.10^{-6}$
90	$0,222.10^{-3}$	$0,476.10^{-4}$	$0,767.10^{-5}$	$0,135.10^{-5}$	$0,168.10^{-6}$
92	$0,270.10^{-3}$	$0,600.10^{-4}$	$0,101.10^{-4}$	$0,183.10^{-5}$	$0,240.10^{-6}$
94	$0,326.10^{-3}$	$0,748.10^{-4}$	$0,131.10^{-4}$	$0,245.10^{-5}$	$0,337.10^{-6}$
96	$0,390.10^{-8}$	$0,924.10^{-4}$	$0,168.10^{-4}$	$0,325.10^{-5}$	$0,466.10^{-6}$

Relative Error in Improved Quadratures

k \ n	16	17	18	19	20
50	$0{,}138.10^{-15}$				
52	$0{,}196.10^{-14}$				
54	$0{,}147.10^{-13}$	$0{,}164.10^{-16}$			
56	$0{,}772.10^{-13}$	$0{,}244.10^{-15}$	$0{,}182.10^{-17}$		
58	$0{,}320.10^{-12}$	$0{,}192.10^{-14}$	$0{,}285.10^{-16}$		
60	$0{,}111.10^{-11}$	$0{,}105.10^{-13}$	$0{,}284.10^{-15}$	$0{,}216.10^{-18}$	
62	$0{,}337.10^{-11}$	$0{,}455.10^{-13}$	$0{,}135.10^{-14}$	$0{,}354.10^{-17}$	$0{,}243.10^{-19}$
64	$0{,}915.10^{-11}$	$0{,}165.10^{-12}$	$0{,}607.10^{-14}$	$0{,}304.10^{-16}$	$0{,}417.10^{-18}$
66	$0{,}227.10^{-10}$	$0{,}520.10^{-12}$	$0{,}229.10^{-13}$	$0{,}182.10^{-15}$	$0{,}874.10^{-17}$
68	$0{,}521.10^{-10}$	$0{,}146.10^{-11}$	$0{,}750.10^{-13}$	$0{,}856.10^{-15}$	$0{,}284.10^{-16}$
70	$0{,}112.10^{-9}$	$0{,}376.10^{-11}$	$0{,}219.10^{-12}$	$0{,}885.10^{-14}$	$0{,}114.10^{-15}$
72	$0{,}228.10^{-9}$	$0{,}894.10^{-11}$	$0{,}588.10^{-12}$	$0{,}114.10^{-13}$	$0{,}463.10^{-15}$
74	$0{,}442.10^{-9}$	$0{,}199.10^{-10}$	$0{,}148.10^{-11}$	$0{,}345.10^{-13}$	$0{,}163.10^{-14}$
76	$0{,}818.10^{-9}$	$0{,}417.10^{-10}$	$0{,}329.10^{-11}$	$0{,}948.10^{-13}$	$0{,}511.10^{-14}$
78	$0{,}146.10^{-8}$	$0{,}831.10^{-10}$	$0{,}711.10^{-11}$	$0{,}240.10^{-12}$	$0{,}145.10^{-13}$
80	$0{,}251.10^{-8}$	$0{,}158.10^{-9}$	$0{,}146.10^{-10}$	$0{,}569.10^{-12}$	$0{,}380.10^{-13}$
82	$0{,}419.10^{-8}$	$0{,}290.10^{-9}$	$0{,}286.10^{-10}$	$0{,}127.10^{-11}$	$0{,}928.10^{-13}$
84	$0{,}679.10^{-8}$	$0{,}512.10^{-9}$	$0{,}538.10^{-10}$	$0{,}268.10^{-11}$	$0{,}213.10^{-12}$
86	$0{,}107.10^{-7}$	$0{,}876.10^{-9}$	$0{,}976.10^{-10}$	$0{,}589.10^{-11}$	$0{,}462.10^{-12}$
88	$0{,}166.10^{-7}$	$0{,}146.10^{-8}$	$0{,}171.10^{-9}$	$0{,}104.10^{-10}$	$0{,}956.10^{-12}$
90	$0{,}251.10^{-7}$	$0{,}235.10^{-8}$	$0{,}291.10^{-9}$	$0{,}193.10^{-10}$	$0{,}189.10^{-11}$
92	$0{,}372.10^{-7}$	$0{,}372.10^{-8}$	$0{,}482.10^{-9}$	$0{,}348.10^{-10}$	$0{,}361.10^{-11}$
94	$0{,}541.10^{-7}$	$0{,}574.10^{-8}$	$0{,}777.10^{-9}$	$0{,}605.10^{-10}$	$0{,}665.10^{-11}$
96	$0{,}775.10^{-7}$	$0{,}869.10^{-8}$	$0{,}123.10^{-8}$	$0{,}102.10^{-9}$	$0{,}118.10^{-10}$
98	$0{,}109.10^{-6}$	$0{,}129.10^{-7}$	$0{,}189.10^{-8}$	$0{,}169.10^{-9}$	$0{,}205.10^{-10}$
100	$0{,}152.10^{-6}$	$0{,}188.10^{-7}$	$0{,}287.10^{-8}$	$0{,}272.10^{-9}$	$0{,}346.10^{-10}$
102	$0{,}208.10^{-6}$	$0{,}270.10^{-7}$	$0{,}426.10^{-8}$	$0{,}429.10^{-9}$	$0{,}569.10^{-10}$
104	$0{,}282.10^{-6}$	$0{,}382.10^{-7}$	$0{,}623.10^{-8}$	$0{,}662.10^{-9}$	$0{,}915.10^{-10}$
106	$0{,}378.10^{-6}$	$0{,}533.10^{-7}$	$0{,}897.10^{-8}$	$0{,}100.10^{-8}$	$0{,}144.10^{-9}$
108	$0{,}500.10^{-6}$	$0{,}734.10^{-7}$	$0{,}127.10^{-7}$	$0{,}149.10^{-8}$	$0{,}222.10^{-9}$
110	$0{,}655.10^{-6}$	$0{,}998.10^{-7}$	$0{,}178.10^{-7}$	$0{,}219.10^{-8}$	$0{,}337.10^{-9}$
112	$0{,}850.10^{-6}$	$0{,}134.10^{-6}$	$0{,}246.10^{-7}$	$0{,}816.10^{-8}$	$0{,}503.10^{-9}$
114	$0{,}109.10^{-5}$	$0{,}178.10^{-6}$	$0{,}336.10^{-7}$	$0{,}450.10^{-8}$	$0{,}739.10^{-9}$
116	$0{,}139.10^{-5}$	$0{,}235.10^{-6}$	$0{,}454.10^{-7}$	$0{,}631.10^{-8}$	$0{,}107.10^{-8}$
118	$0{,}176.10^{-5}$	$0{,}306.10^{-6}$	$0{,}607.10^{-7}$	$0{,}876.10^{-8}$	$0{,}153.10^{-8}$
120	$0{,}221.10^{-5}$	$0{,}396.10^{-6}$	$0{,}802.10^{-7}$	$0{,}120.10^{-7}$	$0{,}215.10^{-8}$
122	$0{,}276.10^{-5}$	$0{,}508.10^{-6}$	$0{,}105.10^{-6}$	$0{,}163.10^{-7}$	$0{,}299.10^{-8}$
124	$0{,}341.10^{-5}$	$0{,}646.10^{-6}$	$0{,}137.10^{-6}$	$0{,}218.10^{-7}$	$0{,}412.10^{-8}$
126	$0{,}419.10^{-5}$	$0{,}814.10^{-6}$	$0{,}176.10^{-6}$	$0{,}290.10^{-7}$	$0{,}560.10^{-8}$
128	$0{,}511.10^{-5}$	$0{,}102.10^{-5}$	$0{,}225.10^{-6}$	$0{,}382.10^{-7}$	$0{,}755.10^{-8}$

Relative Error in Improved Quadratures

k \ n	21	22	23	24	25
66	$0{,}290 \cdot 10^{-20}$				
68	$0{,}519 \cdot 10^{-19}$	$0{,}330 \cdot 10^{-21}$			
70	$0{,}484 \cdot 10^{-18}$	$0{,}615 \cdot 10^{-20}$			
72	$0{,}314 \cdot 10^{-17}$	$0{,}598 \cdot 10^{-19}$	$0{,}394 \cdot 10^{-22}$		
74	$0{,}159 \cdot 10^{-16}$	$0{,}403 \cdot 10^{-18}$	$0{,}765 \cdot 10^{-21}$	$0{,}453 \cdot 10^{-23}$	
76	$0{,}670 \cdot 10^{-16}$	$0{,}212 \cdot 10^{-17}$	$0{,}772 \cdot 10^{-20}$	$0{,}913 \cdot 10^{-22}$	
78	$0{,}244 \cdot 10^{-15}$	$0{,}922 \cdot 10^{-17}$	$0{,}539 \cdot 10^{-19}$	$0{,}955 \cdot 10^{-21}$	$0{,}543 \cdot 10^{-24}$
80	$0{,}790 \cdot 10^{-15}$	$0{,}347 \cdot 10^{-16}$	$0{,}293 \cdot 10^{-18}$	$0{,}691 \cdot 10^{-20}$	$0{,}114 \cdot 10^{-22}$
82	$0{,}282 \cdot 10^{-14}$	$0{,}116 \cdot 10^{-15}$	$0{,}132 \cdot 10^{-17}$	$0{,}389 \cdot 10^{-19}$	$0{,}123 \cdot 10^{-21}$
84	$0{,}626 \cdot 10^{-14}$	$0{,}351 \cdot 10^{-15}$	$0{,}514 \cdot 10^{-17}$	$0{,}181 \cdot 10^{-18}$	$0{,}921 \cdot 10^{-21}$
86	$0{,}157 \cdot 10^{-13}$	$0{,}977 \cdot 10^{-15}$	$0{,}177 \cdot 10^{-16}$	$0{,}726 \cdot 10^{-18}$	$0{,}535 \cdot 10^{-20}$
88	$0{,}371 \cdot 10^{-13}$	$0{,}253 \cdot 10^{-14}$	$0{,}552 \cdot 10^{-16}$	$0{,}258 \cdot 10^{-17}$	$0{,}257 \cdot 10^{-19}$
90	$0{,}827 \cdot 10^{-13}$	$0{,}613 \cdot 10^{-14}$	$0{,}158 \cdot 10^{-15}$	$0{,}829 \cdot 10^{-17}$	$0{,}106 \cdot 10^{-18}$
92	$0{,}176 \cdot 10^{-12}$	$0{,}140 \cdot 10^{-13}$	$0{,}420 \cdot 10^{-15}$	$0{,}244 \cdot 10^{-16}$	$0{,}389 \cdot 10^{-18}$
94	$0{,}856 \cdot 10^{-14}$	$0{,}306 \cdot 10^{-13}$	$0{,}105 \cdot 10^{-14}$	$0{,}666 \cdot 10^{-16}$	$0{,}128 \cdot 10^{-17}$
96	$0{,}697 \cdot 10^{-12}$	$0{,}638 \cdot 10^{-13}$	$0{,}246 \cdot 10^{-14}$	$0{,}170 \cdot 10^{-15}$	$0{,}388 \cdot 10^{-17}$
98	$0{,}181 \cdot 10^{-11}$	$0{,}128 \cdot 10^{-12}$	$0{,}550 \cdot 10^{-14}$	$0{,}411 \cdot 10^{-15}$	$0{,}109 \cdot 10^{-16}$
100	$0{,}240 \cdot 10^{-11}$	$0{,}246 \cdot 10^{-12}$	$0{,}117 \cdot 10^{-13}$	$0{,}942 \cdot 10^{-15}$	$0{,}286 \cdot 10^{-16}$
102	$0{,}424 \cdot 10^{-11}$	$0{,}459 \cdot 10^{-12}$	$0{,}240 \cdot 10^{-13}$	$0{,}206 \cdot 10^{-14}$	$0{,}706 \cdot 10^{-16}$
104	$0{,}730 \cdot 10^{-11}$	$0{,}831 \cdot 10^{-12}$	$0{,}474 \cdot 10^{-13}$	$0{,}432 \cdot 10^{-14}$	$0{,}166 \cdot 10^{-15}$
106	$0{,}123 \cdot 10^{-10}$	$0{,}146 \cdot 10^{-11}$	$0{,}904 \cdot 10^{-13}$	$0{,}872 \cdot 10^{-14}$	$0{,}371 \cdot 10^{-15}$
108	$0{,}201 \cdot 10^{-10}$	$0{,}250 \cdot 10^{-11}$	$0{,}167 \cdot 10^{-12}$	$0{,}170 \cdot 10^{-13}$	$0{,}796 \cdot 10^{-15}$
110	$0{,}328 \cdot 10^{-10}$	$0{,}419 \cdot 10^{-11}$	$0{,}300 \cdot 10^{-12}$	$0{,}321 \cdot 10^{-13}$	$0{,}164 \cdot 10^{-14}$
112	$0{,}508 \cdot 10^{-10}$	$0{,}686 \cdot 10^{-11}$	$0{,}525 \cdot 10^{-12}$	$0{,}588 \cdot 10^{-13}$	$0{,}327 \cdot 10^{-14}$
114	$0{,}785 \cdot 10^{-10}$	$0{,}110 \cdot 10^{-10}$	$0{,}895 \cdot 10^{-12}$	$0{,}105 \cdot 10^{-12}$	$0{,}630 \cdot 10^{-14}$
116	$0{,}119 \cdot 10^{-9}$	$0{,}173 \cdot 10^{-10}$	$0{,}149 \cdot 10^{-11}$	$0{,}182 \cdot 10^{-12}$	$0{,}118 \cdot 10^{-13}$
118	$0{,}178 \cdot 10^{-9}$	$0{,}267 \cdot 10^{-10}$	$0{,}244 \cdot 10^{-11}$	$0{,}310 \cdot 10^{-12}$	$0{,}214 \cdot 10^{-13}$
120	$0{,}262 \cdot 10^{-9}$	$0{,}406 \cdot 10^{-10}$	$0{,}391 \cdot 10^{-11}$	$0{,}516 \cdot 10^{-12}$	$0{,}380 \cdot 10^{-13}$
122	$0{,}380 \cdot 10^{-9}$	$0{,}608 \cdot 10^{-10}$	$0{,}614 \cdot 10^{-11}$	$0{,}841 \cdot 10^{-12}$	$0{,}658 \cdot 10^{-13}$
124	$0{,}548 \cdot 10^{-9}$	$0{,}896 \cdot 10^{-10}$	$0{,}949 \cdot 10^{-11}$	$0{,}135 \cdot 10^{-11}$	$0{,}111 \cdot 10^{-12}$
126	$0{,}767 \cdot 10^{-9}$	$0{,}130 \cdot 10^{-9}$	$0{,}144 \cdot 10^{-10}$	$0{,}212 \cdot 10^{-11}$	$0{,}185 \cdot 10^{-12}$
128	$0{,}107 \cdot 10^{-8}$	$0{,}187 \cdot 10^{-9}$	$0{,}216 \cdot 10^{-10}$	$0{,}327 \cdot 10^{-11}$	$0{,}301 \cdot 10^{-12}$

Relative Error in Improved Quadratures

k \ n	26	27	28	29	30
80	$0,628.10^{-25}$				
82	$0,136.10^{-23}$				
84	$0,153.10^{-22}$	$0,756.10^{-26}$			
86	$0,118.10^{-21}$	$0,169.10^{-24}$	$0,880.10^{-27}$		
88	$0,707.10^{-21}$	$0,196.10^{-23}$	$0\ 204.10^{-25}$		
90	$0,350.10^{-20}$	$0,157.10^{-22}$	$0,244.10^{-24}$	$0,106.10^{-27}$	
92	$0,149.10^{-19}$	$0,967.10^{-22}$	$0,200.10^{-23}$	$0,253.10^{-26}$	$0,124.10^{-28}$
94	$0,562.10^{-19}$	$0,494.10^{-21}$	$0,128.10^{-22}$	$0,313.10^{-25}$	$0,306.10^{-27}$
96	$0,191.10^{-18}$	$0,216.10^{-20}$	$0,671.10^{-22}$	$0,265.10^{-24}$	$0,389.10^{-26}$
98	$0,593.10^{-18}$	$0,838.10^{-20}$	$0,302.10^{-21}$	$0,174.10^{-23}$	$0,339.10^{-25}$
100	$0,171.10^{-17}$	$0,292.10^{-19}$	$0,120.10^{-20}$	$0,938.10^{-23}$	$0,229.10^{-24}$
102	$0,459.10^{-17}$	$0,981.10^{-19}$	$0,480.10^{-20}$	$0,434.10^{-22}$	$0,127.10^{-23}$
104	$0,116.10^{-16}$	$0,275.10^{-18}$	$0,141.10^{-19}$	$0,177.10^{-21}$	$0,604.10^{-23}$
106	$0,280.10^{-16}$	$0,758.10^{-18}$	$0,426.10^{-19}$	$0,651.10^{-21}$	$0,253.10^{-22}$
108	$0,641.10^{-16}$	$0,197.10^{-17}$	$0,120.10^{-18}$	$0,218.10^{-20}$	$0,953.10^{-22}$
110	$0,140.10^{-15}$	$0,483.10^{-17}$	$0,320.10^{-18}$	$0,677.10^{-20}$	$0,327.10^{-21}$
112	$0,296.10^{-15}$	$0,113.10^{-16}$	$0,804.10^{-18}$	$0,196.10^{-19}$	$0,104.10^{-20}$
114	$0,602.10^{-15}$	$0,254.10^{-16}$	$0,192.10^{-17}$	$0,532.10^{-19}$	$0,307.10^{-20}$
116	$0,118.10^{-14}$	$0,546.10^{-16}$	$0,441.10^{-17}$	$0,137.10^{-18}$	$0,854.10^{-20}$
118	$0,226.10^{-14}$	$0,113.10^{-15}$	$0,969.10^{-17}$	$0,334.10^{-18}$	$0,224.10^{-19}$
120	$0,419.10^{-14}$	$0,227.10^{-15}$	$0,205.10^{-16}$	$0,782.10^{-18}$	$0,560.10^{-19}$
122	$0,757.10^{-14}$	$0,442.10^{-15}$	$0,420.10^{-16}$	$0,175.10^{-17}$	$0,134.10^{-18}$
124	$0,134.10^{-13}$	$0,836.10^{-15}$	$0,833.10^{-16}$	$0,379.10^{-17}$	$0,306.10^{-18}$
126	$0,230.10^{-13}$	$0,153.10^{-14}$	$0,160.10^{-15}$	$0,790.10^{-17}$	$0,675.10^{-18}$
128	$0,389.10^{-13}$	$0,276.10^{-14}$	$0,301.10^{-15}$	$0,160.10^{-16}$	$0,144.10^{-17}$

Relative Error in Improved Quadratures

k \ n	31	32	33	34	35
96	$0,150.10^{-29}$				
98	$0,381.10^{-28}$	$0,176.10^{-30}$			
100	$0,498.10^{-27}$	$0,461.10^{-29}$			
102	$0,447.10^{-26}$	$0,620.10^{-28}$	$0,213.10^{-31}$		
104	$0,310.10^{-25}$	$0,572.10^{-27}$	$0,573.10^{-30}$	$0,252.10^{-32}$	
106	$0,176.10^{-24}$	$0,407.10^{-26}$	$0,798.10^{-29}$	$0,695.10^{-31}$	
108	$0,861.10^{-24}$	$0,237.10^{-25}$	$0,752.10^{-28}$	$0,988.10^{-30}$	$0,305.10^{-33}$
110	$0,370.10^{-23}$	$0,119.10^{-24}$	$0,549.10^{-27}$	$0,961.10^{-29}$	$0,866.10^{-32}$
112	$0,143.10^{-22}$	$0,524.10^{-24}$	$0,329.10^{-26}$	$0,720.10^{-28}$	$0,126.10^{-30}$
114	$0,502.10^{-22}$	$0,207.10^{-23}$	$0,169.10^{-25}$	$0,442.10^{-27}$	$0,126.10^{-29}$
116	$0,163.10^{-21}$	$0,746.10^{-23}$	$0,761.10^{-25}$	$0,282.10^{-26}$	$0,966.10^{-29}$
118	$0,493.10^{-21}$	$0,248.10^{-22}$	$0,308.10^{-24}$	$0,107.10^{-25}$	$0,608.10^{-28}$
120	$0,140.10^{-20}$	$0,766.10^{-22}$	$0,113.10^{-23}$	$0,444.10^{-25}$	$0,327.10^{-27}$
122	$0,376.10^{-20}$	$0,222.10^{-21}$	$0,384.10^{-23}$	$0,167.10^{-24}$	$0,154.10^{-26}$
124	$0,958.10^{-20}$	$0,609.10^{-21}$	$0,121.10^{-22}$	$0,579.10^{-24}$	$0,653.10^{-26}$
126	$0,233.10^{-19}$	$0,159.10^{-20}$	$0,360.10^{-22}$	$0,187.10^{-23}$	$0,251.10^{-25}$
128	$0,545.10^{-19}$	$0,394.10^{-20}$	$0,101.10^{-21}$	$0,566.10^{-23}$	$0,890.10^{-25}$
130	$0,122.10^{-18}$	$0,938.10^{-20}$	$0,267.10^{-21}$	$0,161.10^{-22}$	$0,293.10^{-24}$
132	$0,265.10^{-18}$	$0,215.10^{-19}$	$0,677.10^{-21}$	$0,437.10^{-22}$	$0,905.10^{-24}$
134	$0,556.10^{-18}$	$0,474.10^{-19}$	$0,164.10^{-20}$	$0,113.10^{-21}$	$0,263.10^{-23}$
136	$0,113.10^{-17}$	$0,101.10^{-18}$	$0,383.10^{-20}$	$0,279.10^{-21}$	$0,727.10^{-23}$
138	$0,224.10^{-17}$	$0,210.10^{-18}$	$0,861.10^{-20}$	$0,663.10^{-21}$	$0,191.10^{-22}$
140	$0,430.10^{-17}$	$0,421.10^{-18}$	$0,187.10^{-19}$	$0,152.10^{-20}$	$0,482.10^{-22}$
142	$0,809.10^{-17}$	$0,825.10^{-18}$	$0,394.10^{-19}$	$0,335.10^{-20}$	$0,116.10^{-21}$
144	$0,148.10^{-16}$	$0,158.10^{-17}$	$0,806.10^{-19}$	$0,719.10^{-20}$	$0,271.10^{-21}$
146	$0,266.10^{-16}$	$0,294.10^{-17}$	$0,160.10^{-18}$	$0,150.10^{-19}$	$0,610.10^{-21}$
148	$0,469.10^{-16}$	$0,536.10^{-17}$	$0,811.10^{-18}$	$0,302.10^{-19}$	$0,133.10^{-20}$
150	$0,809.10^{-16}$	$0,958.10^{-17}$	$0,590.10^{-18}$	$0,597.10^{-19}$	$0,281.10^{-20}$
152	$0,137.10^{-15}$	$0,168.10^{-16}$	$0,109.10^{-17}$	$0,115.10^{-18}$	$0,578.10^{-20}$
154	$0,229.10^{-15}$	$0,289.10^{-16}$	$0,198.10^{-17}$	$0,216.10^{-18}$	$0,116.10^{-19}$
156	$0,375.10^{-15}$	$0,488.10^{-16}$	$0,358.10^{-17}$	$0,398.10^{-18}$	$0,226.10^{-19}$
158	$0,605.10^{-15}$	$0,812.10^{-16}$	$0,616.10^{-17}$	$0,718.10^{-18}$	$0,482.10^{-19}$
160	$0,962.10^{-15}$	$0,133.10^{-15}$	$0,106.10^{-16}$	$0,127.10^{-17}$	$0,808.10^{-19}$

Relative Error in Improved Quadratures

n \ k	36	37	38	39	40
110	$0{,}361.10^{-34}$				
112	$0{,}105.10^{-32}$				
114	$0{,}157.10^{-31}$	$0{,}488.10^{-35}$			
116	$0{,}161.10^{-30}$	$0{,}131.10^{-33}$	$0{,}521.10^{-36}$		
118	$0{,}126.10^{-29}$	$0{,}201.10^{-32}$	$0{,}159.10^{-34}$		
120	$0{,}815.10^{-29}$	$0{,}210.10^{-31}$	$0{,}250.10^{-33}$	$0{,}633.10^{-37}$	
122	$0{,}449.10^{-28}$	$0{,}169.10^{-30}$	$0{,}269.10^{-32}$	$0{,}199.10^{-35}$	$0{,}659.10^{-38}$
124	$0{,}216.10^{-27}$	$0{,}112.10^{-29}$	$0{,}221.10^{-31}$	$0{,}319.10^{-34}$	$0{,}239.10^{-36}$
126	$0{,}936.10^{-27}$	$0{,}628.10^{-29}$	$0{,}149.10^{-30}$	$0{,}350.10^{-33}$	$0{,}398.10^{-35}$
128	$0{,}368.10^{-26}$	$0{,}310.10^{-28}$	$0{,}858.10^{-30}$	$0{,}295.10^{-32}$	$0{,}447.10^{-34}$
130	$0{,}133.10^{-25}$	$0{,}137.10^{-27}$	$0{,}433.10^{-29}$	$0{,}204.10^{-31}$	$0{,}385.10^{-33}$
132	$0{,}447.10^{-25}$	$0{,}549.10^{-27}$	$0{,}195.10^{-28}$	$0{,}120.10^{-30}$	$0{,}271.10^{-32}$
134	$0{,}141.10^{-24}$	$0{,}203.10^{-26}$	$0{,}799.10^{-28}$	$0{,}615.10^{-30}$	$0{,}163.10^{-31}$
136	$0{,}418.10^{-24}$	$0{,}695.10^{-26}$	$0{,}301.10^{-27}$	$0{,}283.10^{-29}$	$0{,}855.10^{-31}$
138	$0{,}118.10^{-23}$	$0{,}228.10^{-25}$	$0{,}105.10^{-26}$	$0{,}118.10^{-28}$	$0{,}402.10^{-30}$
140	$0{,}315.10^{-23}$	$0{,}674.10^{-25}$	$0{,}344.10^{-26}$	$0{,}454.10^{-28}$	$0{,}171.10^{-29}$
142	$0{,}808.10^{-23}$	$0{,}193.10^{-24}$	$0{,}106.10^{-25}$	$0{,}162.10^{-27}$	$0{,}670.10^{-29}$
144	$0{,}199.10^{-22}$	$0{,}527.10^{-24}$	$0{,}310.10^{-25}$	$0{,}589.10^{-27}$	$0{,}243.10^{-28}$
146	$0{,}471.10^{-22}$	$0{,}138.10^{-23}$	$0{,}860.10^{-25}$	$0{,}169.10^{-26}$	$0{,}826.10^{-28}$
148	$0{,}108.10^{-21}$	$0{,}345.10^{-23}$	$0{,}228.10^{-24}$	$0{,}508.10^{-26}$	$0{,}264.10^{-27}$
150	$0{,}239.10^{-21}$	$0{,}831.10^{-23}$	$0{,}582.10^{-24}$	$0{,}142.10^{-25}$	$0{,}798.10^{-27}$
152	$0{,}514.10^{-21}$	$0{,}193.10^{-22}$	$0{,}142.10^{-23}$	$0{,}384.10^{-25}$	$0{,}230.10^{-26}$
154	$0{,}107.10^{-20}$	$0{,}435.10^{-22}$	$0{,}337.10^{-23}$	$0{,}995.10^{-25}$	$0{,}631.10^{-26}$
156	$0{,}218.10^{-20}$	$0{,}950.10^{-22}$	$0{,}772.10^{-23}$	$0{,}248.10^{-24}$	$0{,}166.10^{-25}$
158	$0{,}433.10^{-20}$	$0{,}202.10^{-21}$	$0{,}171.10^{-22}$	$0{,}596.10^{-24}$	$0{,}421.10^{-25}$
160	$0{,}840.10^{-20}$	$0{,}417.10^{-21}$	$0{,}369.10^{-22}$	$0{,}139.10^{-23}$	$0{,}103.10^{-24}$

Auxiliary Table of Errors for Gaussian Quadratures

n \ Δ	10^{-1}	10^{-2}	10^{-3}	10^{-4}	10^{-5}	10^{-6}	10^{-7}	10^{-8}	10^{-9}	0
1										0
2										2
3										4
4	8									6
5	12									8
6	18	12								10
7	26	16								12
8	32	20	16							14
9	42	26	20							16
10	50	32	24	20						18
11	62	38	28	24	22					20
12	72	46	34	28	24					22
13	84	54	40	32	28	26				24
14	98	62	46	38	32	28				26
15	112	70	52	42	36	32	30			28
16	128	80	60	48	40	36	32			30
17	144	90	66	54	46	40	36	34		32
18	160	100	74	60	50	44	40	38	36	34
19	178	112	82	66	56	50	44	40	38	36
20	198	124	92	74	62	54	48	44	42	38
21	218	136	100	80	68	60	52	48	44	40
22	238	148	110	88	74	64	58	52	48	42
23	260	162	120	96	80	70	62	56	52	44
24	282	176	130	104	88	76	68	62	56	46
25	306	190	140	112	94	82	72	66	60	48
26	330	206	152	122	102	88	78	70	66	50
27	356	222	164	130	110	94	84	76	70	52
28	382	238	176	140	118	102	90	82	74	54
29	410	256	188	150	126	108	96	86	80	56
30	438	274	202	160	134	116	102	92	84	58
31	468	302	214	170	142	124	108	98	90	60
32	498	310	228	182	152	132	116	104	94	62
33	530	330	242	194	162	140	122	110	100	64
34	562	350	258	204	170	148	130	116	106	66
35	594	370	272	216	180	156	138	124	112	68
36	628	392	288	230	192	164	144	130	118	70
37	664	414	304	242	202	174	152	138	124	72
38	700	436	320	254	212	182	160	144	132	74
39	736	458	338	268	224	192	170	152	138	76
40	774	482	354	282	234	202	178	160	144	78

The table presents the maximum even power k for which the Gaussian quadrature of order n, when integrating x^k on $[-1, +1]$, ensures a relative error not exceeding the magnitude of Δ which appears at the head of the column.

140

Auxiliary Table of Errors for Improved Quadratures

n \ Δ	10^{-1}	10^{-2}	10^{-3}	10^{-4}	10^{-5}	10^{-6}	10^{-7}	10^{-8}	10^{-9}	0
1										4
2	18									6
3	36	14								10
4	62	24	16	14						12
5	92	38	24	20	18					16
6	132	52	34	26	22	20				18
7	174	72	46	36	30	26	24			22
8	226	90	58	44	36	32	28	26		24
9	282	114	74	56	46	40	34	32	30	28
10	346	138	88	68	56	46	40	36	34	30
11	414	168	108	82	66	56	50	44	40	34
12	490	196	126	96	78	66	56	50	46	36
13	570	232	148	112	90	76	66	58	52	40
14	660	266	170	128	104	88	76	66	60	42
15	754	306	194	148	120	100	86	76	68	46
16	856	344	220	166	134	112	96	84	76	48
17	960	390	248	188	152	126	110	96	86	52
18	1076	434	276	210	168	140	120	106	94	54
19	1194	484	308	234	188	156	134	118	104	58
20	1322	534	340	256	206	172	148	128	114	60
21	1452	588	374	284	228	190	162	142	126	64
22	1592	642	410	310	248	208	178	154	138	66
23	1736	702	448	338	272	226	194	170	150	70
24	1888	762	486	366	294	246	210	182	162	72
25	2044	828	528	398	320	266	228	198	176	76
26	2210	892	568	428	344	286	244	214	188	78
27	2378	962	612	462	372	310	264	230	204	82
28	2556	1032	658	496	398	332	282	246	218	84
29	2736	1108	706	532	428	356	304	264	234	88
30	2926	1182	754	568	456	380	324	282	250	90
31	3120	1264	804	606	486	406	346	302	266	94
32	3324	1344	856	644	518	430	368	320	282	96
33	3530	1430	910	686	550	458	390	340	300	100
34	3746	1514	964	726	582	484	414	360	318	102
35	3964	1604	1022	770	618	514	438	382	336	106
36	4190	1694	1078	814	652	542	462	402	356	108
37	4424	1790	1140	858	688	574	488	426	376	112
38	4664	1886	1200	904	726	604	514	446	394	114
39	4908	1986	1264	952	764	636	542	472	416	118
40	5162	2088	1328	1000	802	668	568	494	436	120

The table presents the maximum even power k for which the improved quadrature of order n, when integrating x^k on $[-1, +1]$, ensures a relative error not exceeding the magnitude of Δ which appears at the head of the column.

Order of Integration Ensuring a Given Relative Accuracy for Gaussian Quadratures

Function	Integration interval	Relative accuracy								
		10^{-1}	10^{-2}	10^{-3}	10^{-4}	10^{-5}	10^{-6}	10^{-7}	10^{-8}	10^{-9}
e^x	0;1	1	2	2	3	3	3	4	4	4
	0;2	2	2	3	3	4	4	5	5	5
	0;5	2	3	4	5	5	6	6	7	7
	0;10	3	4	5	6	7	7	8	9	9
	0;20	5	6	7	8	9	10	11	11	12
	0;30	6	7	9	10	11	12	13	14	15
$\sin x$*	0;1	1	2	2	3	3	3	4	4	5
	0;2	2	2	3	3	4	4	5	5	5
	0;5	2	3	4	5	5	6	6	7	7
	0;10	4	5	6	7	7	8	9	9	10
	0;20	7	8	9	10	11	11	12	13	14
	0;30	10	11	12	13	14	15	16	17	17
	0;40	13	14	15	16	17	18	19	20	21
$x^{1/64}$	0;M**	1	1	3	10	30	>40	>40	>40	>40
$x^{1/32}$	0;M	1	1	4	13	39	>40	>40	>40	>40
$x^{1/16}$	0;M	1	2	5	16	>40	>40	>40	>40	>40
$x^{1/8}$	0;M	1	2	7	18	>40	>40	>40	>40	>40
$x^{1/6}$	0;M	1	3	7	18	>40	>40	>40	>40	>40
$x^{1/4}$	0;M	1	3	7	17	>40	>40	>40	>40	>40
$x^{1/3}$	0;M	1	3	6	15	36	>40	>40	>40	>40
$x^{1/2}$	0;M	1	3	5	12	25	>40	>40	>40	>40
$x^{2/3}$	0;M	1	2	4	9	17	34	>40	>40	>40
$x^{3/2}$	0;M	2	2	4	6	7	12	18	29	>40
$x^{5/2}$	0;M	2	2	3	4	5	7	9	13	18
$x^{7/2}$	0;M	2	3	3	3	4	5	7	9	11
$x^{9/2}$	0;M	2	3	3	4	4	5	6	7	9
$x^{4/3}$	0;M	1	2	3	5	8	17	21	35	>40
$x^{7/3}$	0;M	2	2	3	4	5	7	10	14	20
$x^{10/3}$	0;M	2	3	3	3	4	5	6	7	9
$x^{5/4}$	0;M	1	2	3	5	8	14	23	38	>40
$x^{9/4}$	0;M	2	2	3	4	5	7	10	14	20
$x^{13/4}$	0;M	2	2	3	3	4	5	7	9	12
$1:(x^2+10^{-2})$	0;1	3	6	9	10	14	17	19	21	21
$1:(x^2+10^{-4})$	0;1	10	20	29	33	>40	>40	>40	>40	>40

*Accuracy is taken relative to $\int |\sin x|\, dx$.   **$M > 0$ is arbitrary.

Order of Integration Ensuring a Given Relative Accuracy for Improved Quadratures

Function	Integration interval	Relative accuracy								
		10^{-1}	10^{-2}	10^{-3}	10^{-4}	10^{-5}	10^{-6}	10^{-7}	10^{-8}	10^{-9}
e^x	0;I	I	I	I	I	I	I	2	2	2
	0;2	I	I	I	I	2	2	2	3	3
	0;5	I	I	2	2	3	3	3	3	4
	0;I0	I	2	3	3	3	4	4	5	5
	0;20	2	3	4	4	5	5	6	6	7
	0;30	2	3	4	5	6	6	6	7	8
$\sin x$*	0;I	I	I	I	I	I	I	2	2	2
	0;2	I	I	I	I	2	2	2	3	3
	0;5	I	I	2	2	3	3	3	4	4
	0;I0	2	3	3	3	4	4	5	5	5
	0;20	3	4	4	5	5	6	7	7	8
	0;30	5	5	6	7	7	8	9	9	I0
	0;40	6	7	8	8	9	I0	II	II	I2
$x^{1/64}$	0;M**	I	I	I	4	II	35	>40	>40	>40
$x^{1/32}$	0;M	I	I	2	5	I5	>40	>40	>40	>40
$x^{1/16}$	0;M	I	I	2	6	I8	>40	>40	>40	>40
$x^{1/8}$	0;M	I	I	3	7	I9	>40	>40	>40	>40
$x^{1/6}$	0;M	I	I	3	7	20	>40	>40	>40	>40
$x^{1/4}$	0;M	I	I	3	6	I6	>40	>40	>40	>40
$x^{1/3}$	0;M	I	I	2	6	I4	32	>40	>40	>40
$x^{1/2}$	0;M	I	I	2	4	9	20	>40	>40	>40
$x^{2/3}$	0;M	I	I	2	3	6	I3	26	>40	>40
$x^{3/2}$	0;M	I	I	I	2	2	4	6	I0	I6
$x^{5/2}$	0;M	I	I	I	2	2	3	3	5	7
$x^{7/2}$	0;M	I	I	I	I	2	3	3	4	6
$x^{9/2}$	0;M	I	I	I	2	2	2	2	3	4
$x^{4/3}$	0;M	I	I	I	2	3	5	8	I2	20
$x^{7/3}$	0;M	I	I	I	2	2	3	3	3	7
$x^{10/3}$	0;M	I	I	I	I	2	3	3	4	5
$x^{5/4}$	0;M	I	I	I	2	3	5	8	I5	22
$x^{9/4}$	0;M	I	I	I	2	2	3	3	5	7
$x^{13/4}$	0;M	I	I	I	I	2	2	3	4	5
$1:(x^2+10^{-2})$	0;I	I	3	4	6	7	8	I0	II	I2
$1:(x^2+10^{-4})$	0;I	4	8	I3	I8	23	27	30	35	37

*Accuracy is taken relative to $\int |\sin x|\, dx$.   **$M > 0$ is arbitrary.